Lecture Notes in Mathematics   2088

Editors:
J.-M. Morel, Cachan
B. Teissier, Paris

W0193018

For further volumes:
http://www.springer.com/series/304

Ju-Yi Yen • Marc Yor

Local Times and Excursion Theory for Brownian Motion

A Tale of Wiener and Itô Measures

 Springer

Ju-Yi Yen
Department of Mathematical Sciences
University of Cincinnati
Cincinnati, OH, USA

Marc Yor
Laboratoire de Probabilités et Modèles
    Aléatoires
Université Paris VI
Paris CX 05, France

ISBN 978-3-319-01269-8   ISBN 978-3-319-01270-4 (eBook)
DOI 10.1007/978-3-319-01270-4
Springer Cham Heidelberg New York Dordrecht London

Lecture Notes in Mathematics ISSN print edition: 0075-8434
                              ISSN electronic edition: 1617-9692

Library of Congress Control Number: 2013948765

Mathematics Subject Classification (2010): 60J55, 60J57, 60J65, 60G17

Preface

This monograph takes up and completes the volume written by the second author and edited by the University of Caracas (Venezuela), following a course given there in July 1994.

The present monograph consists essentially and naturally of three parts:

Part I presents local times for continuous semimartingales, while Part II is devoted to Excursion Theory for Brownian paths, and Part III to some applications of this theory. Chapter 1 gathers some facts which will be helpful throughout the volume.

However, this monograph differs from the Caracas volume in an essential way: the "credo" of the Caracas volume was that, once one knows "something" about the Wiener measure $\mathbf{W}$, one should be able to translate this "something" in terms of Itô's characteristic measure, usually denoted by $\mathbf{n}$, for Brownian excursions.

The union of the translations of these "something," such as the law under $\mathbf{n}(d\varepsilon)$ of the lifetime $V(\varepsilon)$ and/or the height $M(\varepsilon)$ of the generic excursion ε, should provide a full understanding of $\mathbf{n}(d\varepsilon)$ which, in turn, should enrich our understanding of Wiener measure $\mathbf{W}(dw)$.

In fact, as the second author experienced it, when teaching year after year this "local times—excursion" course: the reality is somewhat more complicated, it is true that D. Williams' path decomposition of the excursion straddling the time $T_a = \inf\{t : B_t = a\}$ translates easily into Itô's measure being disintegrated at the maximum of the height of the generic excursion, but, on the other hand, there is a priori no direct way to show that the normalized standard excursion, straddling deterministic time t, say, is equal (in law) to a BES(3) bridge. In fact, this result will follow from the disintegration of Itô's measure $\mathbf{n}$, with respect to its lifetime V.

It is this difficulty which led us to try and present an as easy as possible approach to both measures $\mathbf{W}$ and $\mathbf{n}$, the fine descriptions of which being extremely intricate.

This central point ("the core of the course") being explained, we refer the reader to the remaining plan of this volume, which is self-explanatory. Most proofs are self-contained, and references for the missing points are clearly indicated. For ease of the reader, each chapter has its own set of references, while general references are gathered at the end of the book, together with an index of terms.

Important Note: We consider these references, either at the end of a chapter, or at the end of the book, to convey some essential information for the reader. They may, or may not be cited in the text, but the reader is expected to consider them as a fertile source of material.

Beside this text, there exists a related one, by B. Mallein and the second author, where the same general thread is followed, but in a quite different manner: they only quote—without proofs—the main theorems of the different chapters, while the main body of each chapter consists in a number of exercises which were given, over the years, in exams related to this course.

This second volume may be a useful companion to the present one, with which the reader may hone his/her skills.

M.Y. is very grateful for the support of the von Humboldt–Stiftung during his stay in Freiburg, during the academic year 2011–2012. J.-Y.Y. warmly thanks the Academia Sinica Institute of Mathematics in Taipei, Taiwan, for their hospitality and support during several extended visits. This work was supported in part by NSF Grant DMS-0907513.

Cincinnati, OH, USA   Ju-Yi Yen
Taipei, Taiwan   Marc Yor
Paris, France
April 27, 2013

Contents

Chapter 1
Prerequisites

In this chapter, ten notions or results are gathered, which we assume as background for the remainder of this monograph.

1.1 Brownian Motion

It is not difficult to show the existence of a probability space on which one can construct a Gaussian family $\{B(f); f \in L^2(\mathbb{R}_+, dt)\}$, such that

$$\text{(i)} \ \ E[B(f)] = 0; \qquad \text{(ii)} \ \ E[(B(f))^2] = \int f^2(t)dt.$$

Indeed, from a functional viewpoint, B is a Hilbert space isomorphism

$$B : L^2(\mathbb{R}_+, dt) \rightarrow \mathcal{B}(\subset L^2(\Omega))$$
$$f \rightarrow B(f)$$

so that:

$$B(f) = \sum_{n \geq 1}(f, e_n)_{L^2}G_n$$

where $(e_n; n \geq 1)$ is an orthonormal basis of $L^2(\mathbb{R}_+, dt)$, and $(G_n; n \geq 1)$ is a sequence of centered, reduced independent Gaussian variables.

We shall call Brownian motion, BM in brief, a continuous modification $\{B_t, t \geq 0\}$ of the Gaussian family

$$(B(\mathbf{1}_{[0,t]}); t \geq 0)$$

J.-Y. Yen and M. Yor, *Local Times and Excursion Theory for Brownian Motion*,
Lecture Notes in Mathematics 2088, DOI 10.1007/978-3-319-01270-4_1,
© Springer International Publishing Switzerland 2013

Once the existence of this continuous modification is established (using Kolmogorov's continuity criterion; see Sect. 1.10), it is natural to use (Wiener) integral notation

$$\int_0^\infty f(t)dB_t$$

instead of $B(f)$ since, in the particular case

$$f(t) = \sum \lambda_i \mathbf{1}_{(t_i,t_{i+1}]}(t)$$

one has

$$B(f) = \sum \lambda_i (B_{t_{i+1}} - B_{t_i})$$

1.2 Some Extensions

Given any measurable space $(T, \mathcal{T})$ equipped with a positive σ-finite measure μ, one can, just as in the previous case, define a so-called Gaussian measure $(B(f) \equiv \int f(t)B(\mu(dt)); f \in L^2(T, \mathcal{T}; \mu)$ such that

(i) $E[B(f)] = 0;$ (ii) $E[(B(f))^2] = \int f^2(t)\mu(dt).$

The Brownian sheet corresponds to $T = \mathbb{R}_+^2$ (more generally, $\mathbb{R}_+^n$) and $\mu(dsdt) = dsdt$ the Lebesgue measure.

One can also construct important Gaussian families from a Gaussian measure, by considering:

$$\int \Phi(t, s) B(\mu(ds)).$$

The Lévy's n-parameter Brownian motions and fractional Brownian motions may be defined in this way.

1.3 BM as a Continuous Martingale

The following theorem presents Brownian motion as a prototype for continuous martingales.

Theorem 1.3.1 (Dubins–Schwarz). *Let M be a continuous local martingale such that $M_0 = 0$ and $\langle M \rangle_\infty = \infty$. There exists a BM $(B_u; u \geq 0)$ such that*

$$M_t = B_{\langle M \rangle_t}.$$

Next, here is a partial extension of the preceding theorem to multidimensional continuous martingales.

Theorem 1.3.2 (Knight). *Let $M^{(1)}, M^{(2)}, \ldots, M^{(k)}$ be k continuous local martingales with $M_0^{(i)} = 0$, $\langle M^{(i)} \rangle_\infty = \infty$ and $\langle M^{(i)}, M^{(j)} \rangle_t = 0$ for $i \neq j$; then there exist k independent BM's $(B_u^{(i)}; u \geq 0)$, $i = 1, \ldots, k$ such that*

$$M_t^{(i)} = B_{\langle M^{(i)} \rangle_t}^{(i)}.$$

If moreover $\langle M^{(i)} \rangle_t \equiv \langle M \rangle_t$ for $i = 1, \ldots, k$, i.e., there is a common time change, Theorem 1.3.2 implies that $M_t = B_{\langle M \rangle_t}$ where $B = (B^{(1)}, \ldots, B^{(k)})$ and $B^{(i)}$ are independent BM's. Such multidimensional martingales are called *conformal martingales* (in particular in the case $k = 2$).

Examples of Conformal Martingales.

Let $Z_t \equiv B_t^{(1)} + iB_t^{(2)}$ be a complex BM. If $f \in \mathcal{H}(\mathbb{C})$ is an entire function, which is not constant, then $(M_t = f(Z_t); t \geq 0)$ is a conformal (local) martingale. Then

$$\langle M \rangle_t = \int_0^t ds |f'(Z_s)|^2$$

and Theorem 1.3.2 implies that there exists a $\mathbb{C}$ valued BM $(\hat{Z}_u; u \geq 0)$ such that

$$M_t = \hat{Z}_{\int_0^t ds |f'(Z_s)|^2}.$$

In a general case (i.e. $f \in \mathcal{C}^2(\mathbb{R}^2)$), Itô's "complex" formula may be written as:

$$f(Z_t) = f(Z_0) + \int_0^t \frac{\partial f}{\partial z}(Z_s) dZ_s + \int_0^t \frac{\partial f}{\partial \bar{z}}(Z_s) d\bar{Z}_s + \int_0^t \frac{\partial^2 f}{\partial z \partial \bar{z}}(Z_s) d\langle Z, \bar{Z} \rangle_s$$

and if f is holomorphic, then:

$$f(Z_t) = f(Z_0) + \int_0^t f'(Z_s) dZ_s.$$

More generally again, let $X = M + V$ be a continuous semimartingale in $\mathbb{R}^n$ and $f \in C^2(\mathbb{R}^n)$; then Itô's formula is

$$f(X_t) = f(X_0) + \int_0^t (\nabla f)(X_s) \cdot dX_s + \frac{1}{2} \int_0^t \sum_{i,j} \frac{\partial^2 f}{\partial x_i \partial x_j}(X_s) d \langle X^{(i)}, X^{(j)} \rangle_s.$$

For a detailed exposition see [4].

1.4   Girsanov's Theorem

This fundamental theorem often allows to extend theorems known to be valid for BM to "mild perturbations of BM".

On the canonical space $C(\mathbb{R}_+, \mathbb{R})$, we consider the canonical process $X_t(\omega) = \omega(t)$ and the canonical filtration $\mathcal{F}_t \equiv \sigma\{X_s; s \leq t\}$.

For every $x \in \mathbb{R}$, $\mathbf{W}_x$ will denote the Wiener measure on $\mathcal{F}_\infty$ such that $\mathbf{W}_x(X_0 = x) = 1$.

We shall say that a process Y is a mild perturbation of BM if its law P_Y has the same null sets as $\mathbf{W}$ on each σ-field $\mathcal{F}_t$, i.e. the measure P_Y is such that

$$P_Y|_{\mathcal{F}_t} \sim \mathbf{W}|_{\mathcal{F}_t}; \quad t \geq 0.$$

Example 1.4.1.

(a) Brownian motion with drift μ.

Let $B_t^{(\mu)} = B_t + \mu t$, $t \geq 0$; then the associated measure $\mathbf{W}^{(\mu)}$ satisfies

$$\mathbf{W}^{(\mu)}|_{\mathcal{F}_t} = \exp\left(\mu X_t - \frac{\mu^2}{2} t \right) \mathbf{W}|_{\mathcal{F}_t}.$$

(b) The Cameron–Martin formula.

Let $B_t^{(f)} = B_t + \int_0^t ds\, f(s)$ where $f \in L^2_{loc}(\mathbb{R}_+)$; then the corresponding measure $\mathbf{W}^{(f)}$ satisfies

$$\mathbf{W}^{(f)}|_{\mathcal{F}_t} = \exp\left(\int_0^t f(s) dX_s - \frac{1}{2} \int_0^t f^2(s) ds \right) \mathbf{W}|_{\mathcal{F}_t}.$$

(c) Girsanov's formula.

Let $Z_t = B_t + \int_0^t ds\, \varphi(Z_s)$ where φ is a bounded Borel function. The associated measure $P^{(\varphi)}$ satisfies

$$P^{(\varphi)}|_{\mathcal{F}_t} = \exp\left(\int_0^t \varphi(X_s) dX_s - \frac{1}{2} \int_0^t \varphi^2(X_s) ds \right) \mathbf{W}|_{\mathcal{F}_t}.$$

All these examples are particular cases of Girsanov's theorem, of which we now present the continuous martingale version.

Theorem 1.4.2 (Girsanov–Wong–Van Schuppen). *Given a probability measure P and a $(P, \mathcal{F}_t)$-local martingale M such that Q can be defined with the property*

$$Q|_{\mathcal{F}_t} = \exp\left(M_t - \frac{1}{2}\langle M \rangle_t\right) P|_{\mathcal{F}_t}.$$

Then, if N is a $(P, \mathcal{F}_t)$-local martingale, $N_t - \langle N, M \rangle_t$ is a $(Q, \mathcal{F}_t)$-local martingale.

Corollary 1.4.3. *If N is a $(\mathcal{F}_t)$-BM under P, then $(\tilde{N} \equiv N_t - \langle N, M \rangle_t; t \geq 0)$ is a BM under Q.*

Corollary 1.4.3 holds since $\langle \tilde{N} \rangle_t = \langle N \rangle_t = t$.

Example 1.4.4. If $N = M$ then $M_t = \tilde{M}_t + \langle M \rangle_t$ where $(\tilde{M}_t; t \geq 0)$ is a Q-local martingale.

Let us see how Girsanov theorem applies to Example 1.4.1(c). Let $(X_t; t \geq 0)$ be a BM, i.e. a **W**-martingale, then $M_t = \int_0^t \varphi(X_s) dX_s$ is a **W**-local martingale. The theorem implies that $\tilde{X}_t \equiv X_t - \langle X, M \rangle_t$ is a $P^{(\varphi)}$-local martingale, whence $\tilde{X}_t$ is a $P^{(\varphi)}$-BM, since $\langle \tilde{X} \rangle_t = t$.

Note that

$$\langle X, M \rangle_t = \int_0^t \varphi(X_s) ds.$$

The other examples can be treated similarly.

1.5 Brownian Bridge

The Brownian bridge $b = \{b_u; 0 \leq u \leq 1\}$ is defined as the conditioned process $\{(B_u; u \leq 1) | B_1 = 0\}$.

We shall use the fact that $B_t = (B_t - tB_1) + tB_1$ is the orthogonal decomposition of B_t with respect to $L^2(\sigma(B_1))$, since:

$$E[(B_t - tB_1)B_1] = 0.$$

Now, the Gaussian property implies that $(B_t - tB_1; t \leq 1)$ is independent of B_1, hence:

$$(B_t, t \leq 1 | B_1 = y) \stackrel{(\text{law})}{=} (B_t - tB_1 + ty).$$

We can thus represent the bridge between 0 and y during the time interval $[0, 1]$ as

$$(B_t - tB_1 + ty; t \leq 1)$$

and we denote by $\mathbf{W}_{0 \to y}^{(1)}$ the associated measure. In general, $\mathbf{W}_{x \to y}^{(t)}$ denotes the measure associated to the bridge between x and y during the time interval $[0, t]$, which may be realized as

$$\left(x + \left(B_u - \frac{u}{t} B_t \right) + \frac{u}{t} (y - x); u \leq t \right),$$

where $(B_u; u \leq t)$ is a standard BM starting from 0.

Theorem 1.5.1. $\mathbf{W}_{x \to y}^{(t)}$ *is equivalent to* $\mathbf{W}_x$ *on* $\mathcal{F}_s$ *for* $s < t$.

Proof. Let $F_s \geq 0$ be an $\mathcal{F}_s$-measurable functional, then

$$E_x[F_s f(X_t)] = E_x[E_x(F_s | X_t) f(X_t)] = E_x[F_s P_{t-s} f(X_s)]$$

where $(X_t; t \geq 0)$ is a Markov process with semigroup

$$P_t(x, dy) = p_t(x, y) dy.$$

On the other hand,

$$E_x[F_s P_{t-s} f(X_s)] = E_x[F_s \int f(y) p_{t-s}(X_s, y) dy] = \int f(y) E_x[F_s p_{t-s}(X_s, y)] dy$$

and also

$$E_x[E_x[F_s | X_t] f(X_t)] = \int dy f(y) p_t(x, y) E_{x \to y}^{(t)}(F_s)$$

whence

$$E_{x \to y}^{(t)}(F_s) = \frac{E_x[F_s p_{t-s}(X_s, y)]}{p_t(x, y)}.$$

Thus

$$P_{x \to y}^{(t)} \big|_{\mathcal{F}_s} = \frac{p_{t-s}(X_s, y)}{p_t(x, y)} P_x \big|_{\mathcal{F}_s}.$$

If $x = y = 0$, we have

$$P_{0 \to 0}^{(t)} \big|_{\mathcal{F}_s} = \left(\frac{t}{t - s} \right)^{n/2} \exp \left(\frac{-|X_s|^2}{2(t - s)} \right) P_0 \big|_{\mathcal{F}_s}.$$

As a consequence, we can write the canonical decomposition of the standard Brownian bridge (under $P_{0 \to 0}^{(t)}$) as:

$$X_s = B_s - \int_0^s du \frac{X_u}{t-u}, \quad s \le t,$$

where $(B_s, s \le t)$ is a BM under $P_{0 \to 0}^{(t)}$.   □

1.6   The BES(3) Process as a Doob h-Transform of BM

We use the notation $\mathrm{BES}_a(3)$ for the three-dimensional Bessel process starting from a, and $P_a^{(3)}$ for its law.

Using Girsanov theorem (see Sect. 1.4) one can show the following absolute continuity relation

$$P_a^{(3)}|_{\mathcal{F}_t} = \left(\frac{X_{t \wedge T_0}}{a} \right) \mathbf{W}_a|_{\mathcal{F}_t}.$$

As an important consequence, if $f : \mathbb{R}_+ \times \mathbb{R}_+ \to \mathbb{R}_+$ is a harmonic space-time function, then $\left(\frac{1}{X_t} f(t, X_t); t \ge 0 \right)$ is a $(P_a^{(3)}, \mathcal{F}_t)$ local martingale. The absolute continuity relation, or h-process relation, between a BES(3) and BM is a key property to the proof of Williams' time-reversal theorem.

Theorem 1.6.1 (Williams' time reversal). *Let $(B_t; t \le T_1)$ be a BM starting at 0 and considered up to time $T_1 \equiv \inf\{t \ge 0 : B_t = 1\}$, then*

$$(1 - B_{T_1 - t}; t \le T_1) \overset{(\,law\,)}{=} (R_t; t \le \gamma^1)$$

where $(R_t; t \le \gamma^1)$ denotes a BES(3) process starting at 0 considered up to time $\gamma^1 \equiv \sup\{t \ge 0 : R_t = 1\}$.

1.7   The Beta–Gamma Algebra

Let Z_a be a random variable having Gamma density $h_a(t) = \frac{t^{a-1} e^{-t}}{\Gamma(a)}$ on $\mathbb{R}_+$ and $Z_{a,b}$ a variable with Beta density $\tilde{h}_{a,b}(t) = \frac{t^{a-1}(1-t)^{b-1}}{\beta(a,b)}$ on $[0, 1]$.
  If Z_a and Z_b are independent, then

(i) $Z_a + Z_b \overset{(\text{law})}{=} Z_{a+b}$
(ii) $Z_{a,b} \overset{(\text{law})}{=} \frac{Z_a}{Z_a + Z_b}$

  From (i) and (ii), one gets $Z_a \overset{(\text{law})}{=} Z_{a,b} Z_{a+b}$, which implies

(iii) $(Z_a, Z_b) \overset{(\text{law})}{=} Z_{a+b}(Z_{a,b}, 1 - Z_{a,b})$
   As an application of (iii), one can show that

$$(N^2, N'^2) \overset{(\text{law})}{=} 2T(Z, 1 - Z)$$

where N and N' are two independent standard Gaussian r.v.'s, T is an exponential r.v. with parameter 1 and is independent of $Z \overset{(\text{law})}{=} Z_{1/2,1/2}$, a so-called arc-sine variable.

1.8   The Law of the Maximum of a Positive Continuous Local Martingale, Which Converges to 0

The following universal result for such a local martingale is:

$$\sup_{t \geq 0} M_t \overset{(\text{law})}{=} \frac{M_0}{U},$$

where U is uniform and independent from M_0.
   This is a simple consequence of the optional stopping theorem. Precisely:

Lemma 1.8.1. *Let M be a local continuous martingale with $M_0 = a$, $M_t \geq 0$ and $\lim\limits_{t \to \infty} M_t = 0$. Then*

$$\sup_{t \geq 0} M_t \overset{(\text{law})}{=} \frac{a}{U}$$

where U is a uniform variable on $[0, 1]$.

Proof. Let $y > a$, then

$$a = E[M_{T_y}] = yP(T_y < \infty) = yP\left(\sup_{t \geq 0} M_t \geq y\right),$$

thus

$$P\left(\sup_t M_t \geq y\right) = \frac{a}{y} = P\left(\frac{a}{U} \geq y\right).$$

$\square$

Exercise 1.8.2. The aim of this exercise is to show the identity: for $F_t \geq 0$, $\mathcal{F}_t$-measurable

$$E\left[F_t\left(1 - \frac{M_t}{a}\right)^+\right] = E\left[F_t \mathbf{1}_{(g_\infty^{(a)} \leq t)}\right], \tag{1.8.1}$$

where $M_t \geq 0$, is a continuous local martingale, and $M_t \xrightarrow[t \to \infty]{} 0$, and $g_\infty^{(a)} = \sup\{t : M_t = a\}$.

(a) Note that (1.8.1) is equivalent to:

$$P\left(g_\infty^{(a)} \leq t \mid \mathcal{F}_t\right) = \left(1 - \frac{M_t}{a}\right)^+.$$

(b) Deduce (1.8.1) from $\left(g_\infty^{(a)} \leq t\right) = \left(\sup_{u \geq t} M_u \leq a\right)$, then apply Lemma 1.8.1.

1.9   A First Taste of Enlargement Formulae

We are concerned here with the following theorem.

Theorem 1.9.1. *(a)   If $L \equiv \sup\{t : (t, \omega) \in \Gamma\}$, where Γ is a set belonging to the predictable σ-field of $(\mathcal{F}_t)$, a given filtration, then all $(\mathcal{F}_t)$ martingales remain $(\mathcal{F}_t^L)$ semimartingales, where $(\mathcal{F}_t^L \equiv \mathcal{F}_t \vee \sigma(L \wedge t))$ is the smallest filtration containing $(\mathcal{F}_t)$ and making L a stopping time.*

(b)   If we define $Z_t \equiv Z_t^L = P(L > t \mid \mathcal{F}_t)$, then a generic $(\mathcal{F}_t)$ martingale (M_t) becomes a semimartingale in $(\mathcal{F}_t^L)$, with canonical decomposition:

$$M_t = \tilde{M}_t + \int_0^{L \wedge t} \frac{d \langle M, Z^L \rangle_s}{Z_s^L} + \int_L^t \frac{d \langle M, 1 - Z^L \rangle_s}{1 - Z_s^L}.$$

We have assumed the hypothesis:
(CA): every $(\mathcal{F}_t)$ martingale is continuous and, for any $(\mathcal{F}_t)$ stopping time T, $P(L = T) = 0$.

Such formulae shall be useful when we shall enlarge a given filtration with, say: $\Lambda_a = \sup\{t : R_t = a\}$ for some transient process R.

A number of computations of Z^L are presented in [3].

1.10   Kolmogorov's Continuity Criterion

This important lemma allows to construct continuous modification of a process which satisfies a simple criterion.

Theorem 1.10.1. *Let $X = (X_x)_{x \in I}$ be a random process indexed by a bounded interval I of $\mathbb{R}$, and taking values in a complete metric space (M, d). Assume the existence of three reals $p, \epsilon, C > 0$ such that for every $x, y \in I$:*

$$E[(d(X_x, X_y))^p] \leq C |x - y|^{1+\epsilon}.$$

Then, there exists a modification $\tilde{X}$ of this process X whose trajectories are Hölder with exponent α, for any $\alpha \in]0, \frac{\epsilon}{p}[$. This means that for any $\alpha \in]0, \frac{\epsilon}{p}[$, there exists a constant $C_\alpha(\omega)$ such that for all $x, y \in I$:

$$d(\tilde{X}_x(\omega), \tilde{X}_y(\omega)) \leq C_\alpha(\omega)|x - y|^\alpha.$$

In particular, $\tilde{X}$ is a continuous modification of X.

References

1. T. Jeulin, Semi-martingales et grossissement d'une filtration. *Lecture Notes in Mathematics*, vol. 833. (Springer, Berlin, 1980)
2. T. Jeulin, M. Yor, Grossissement de filtrations: exemples et applications. *Lecture Notes in Mathematics*, vol. 1118. (Springer, Berlin, 1985)
3. R. Mansuy, M. Yor, Random times and enlargements of filtrations in a Brownian setting. *Lecture Notes in Mathematics*, vol. 1873. (Springer, Berlin, 2006)
4. D. Revuz, M. Yor, Continuous martingales and Brownian motion. *Grundlehren der Mathematischen Wissenschaften [Fundamental Principles of Mathematical Sciences]*, vol. 293, 3rd edn. (Springer, Berlin, 1999)

Part I
Local Times of Continuous Semimartingales

Chapter 2
The Existence and Regularity of Semimartingale Local Times

In this chapter, the existence and regularity properties of local times associated to a continuous semimartingale are shown. Some martingales involving the local times are constructed. Existence of principal values for Brownian motion is deduced from the regularity of Brownian local times.

2.1 From Itô's Formula to the Occupation Time Formula

Let $(B_t; t \geq 0)$ be a one-dimensional Brownian motion (BM), we prove that:
the application $f \rightarrow \int_0^t ds\, f(B_s(w))$ defines a measure which is absolutely continuous with respect to Lebesgue measure.

For this purpose, we think of $\int_0^t ds\, f(B_s)$ as the last term in Itô's formula applied to $F(B_t)$ where F is a "second order" primitive of f.

More generally, we shall start with a continuous semimartingale $Y = M + V$ and we consider the integral

$$\int_0^t d\langle Y \rangle_s f(Y_s).$$

We assume that f is continuous with compact support, and define:

$$F'(x) = \int_{-\infty}^{x} f(y)dy = \int_{-\infty}^{\infty} dy\, f(y)\mathbf{1}_{(x>y)}$$

$$F(x) = \int_{-\infty}^{x} d\zeta \int_{-\infty}^{\zeta} dy\, f(y) = \int_{-\infty}^{\infty} dy\, f(y) \int_{y}^{x} d\zeta = \int_{-\infty}^{\infty} dy(x-y)^{+} f(y).$$

J.-Y. Yen and M. Yor, *Local Times and Excursion Theory for Brownian Motion*, Lecture Notes in Mathematics 2088, DOI 10.1007/978-3-319-01270-4_2, © Springer International Publishing Switzerland 2013

Using Itô's formula for $F(Y_t)$, we obtain:

$$\int_{-\infty}^{\infty} dyf(y)(Y_t - y)^+ = \int_{-\infty}^{\infty} dyf(y)(Y_0 - y)^+ + \int_{-\infty}^{\infty} dyf(y) \int_0^t dY_s \mathbf{1}_{(Y_s > y)}$$
$$+ \frac{1}{2} \int_0^t d\langle Y \rangle_s f(Y_s).$$

The change of the order of integration concerning the stochastic integral term can be easily justified using a variant of Fubini's theorem. From this we get the density of occupation formula

$$\frac{1}{2} \int_0^t d\langle Y \rangle_s f(Y_s) = \int_{-\infty}^{\infty} dyf(y)\Big(\frac{1}{2} L_t^y(Y)\Big) \tag{2.1.1}$$

where

$$\frac{1}{2} L_t^y(Y) \equiv (Y_t - y)^+ - (Y_0 - y)^+ - \int_0^t dY_s \mathbf{1}_{(Y_s > y)}. \tag{2.1.2}$$

Had we taken for our choice of F':

$$F'(x) = - \int_x^{\infty} dyf(y),$$

we would have obtained:

$$\frac{1}{2} \tilde{L}_t^y(Y) \equiv (Y_t - y)^- - (Y_0 - y)^- + \int_0^t dY_s \mathbf{1}_{(Y_s < y)}. \tag{2.1.3}$$

Since $a = a^+ - a^-$, formulae (2.1.2) and (2.1.3) differ by:

$$\frac{1}{2}\big(L_t^y(Y) - \tilde{L}_t^y(Y)\big) = \int_0^t dY_s \mathbf{1}_{(Y_s = y)}.$$

2.2 Regularity of Occupation Times

Theorem 2.2.1. *There exists a modification of* (L_t^y) *which is jointly continuous in* t *and right-continuous with left-limits in* y, *and the jumps in the* y *variable are given by:*

$$L_t^y - L_t^{y^-} = 2 \int_0^t dY_s \mathbf{1}_{(Y_s = y)} = 2 \int_0^t dV_s \mathbf{1}_{(Y_s = y)}$$

Remark 2.2.2. We have

$$\int_0^t dM_s \mathbf{1}_{(Y_s=y)} = 0,$$

since

$$\int_0^t d\langle M\rangle_s \mathbf{1}_{(Y_s=y)} = 0$$

by the density of occupation formula (2.1.1).

If $f \to \int_0^t dV_s\, f(Y_s)$ is absolutely continuous then $\int_0^t dV_s \mathbf{1}_{(Y_s=y)} = 0$ and there exists a jointly continuous modification of $L_t^y(Y)$ in (t, y).

Proof of the theorem. This relies on an application of Kolmogorov's continuity criterion, (see Sect. 1.10).

Since $\int_0^t dV_s \mathbf{1}_{(Y_s>y)}$ is jointly continuous in t, and càdlàg in y, it will suffice to apply Kolmogorov's criterion to

$$X_y = \Big(X_y(t) = \int_0^t dM_s \mathbf{1}_{(Y_s>y)}; t \ge 0\Big), \text{ considered to take values in } C([0, \infty), \mathbb{R})$$

Indeed, using the Burkholder–Davis–Gundy (B–D–G) inequality, we have

$$E[\sup_{s\le t} |X_x(s) - X_y(s)|^p] \le C_p E\Big[\Big(\int_0^t d\langle M\rangle_s \mathbf{1}_{(x<Y_s<y)}\Big)^{p/2}\Big].$$

The previous term is then

$$= C_p E\Big[\Big(\int_x^y d\xi L_t^\xi(Y)\Big)^{p/2}\Big] = C_p(y-x)^{p/2} E\Big[\Big(\frac{1}{y-x}\int_x^y d\xi L_t^\xi(Y)\Big)^{p/2}\Big]$$

$$\le C_p(y-x)^{p/2} \sup_\xi E[(L_t^\xi(Y))^{p/2}].$$

On the other hand, Tanaka's formula and another application of B–D–G imply

$$E[(L_t^\xi(Y))^{p/2}] \le C_p E\Big[(\langle Y\rangle_\infty)^{p/4} + \Big(\int_0^\infty |dV_s|\Big)^{p/2}\Big] = C_{p,Y} < \infty$$

where the constant $C_{p,Y}$ can be taken to be finite using localization. To finish the proof choose p such that $\epsilon = p/2 - 1 > 0$. □

Example 2.2.3.

(a) Let $Y = M + V$ be a semimartingale such that

$$V_t = \int_0^t v_s d \langle M \rangle_s,$$

with $\int_0^t |v_s| d \langle M \rangle_s < \infty$; then $L_t^y (Y)$ is jointly continuous, since

$$L_t^y (Y) - L_t^{y^-} (Y) = 2 \int_0^t d \langle M \rangle_s v_s \mathbf{1}_{(Y_s = y)} = 0$$

In particular, if $V \equiv 0$, $L_t^y (Y)$ is continuous in (y, t).

(b) Let $a, b \in \mathbb{R}, a \neq b$, and

$$Y_t = a M_t^+ - b M_t^- \equiv \int_0^t dM_s \left(a\mathbf{1}_{(M_s > 0)} + b\mathbf{1}_{(M_s < 0)} \right) + \frac{a - b}{2} l_t,$$

where $l = L^0(M)$.

Since $V_t = \frac{a-b}{2} l_t$, we have: $L_t^0(Y) - L_t^{0^-} (Y) = (a - b) l_t$; thus $L_t^y (Y)$ is not continuous at $y = 0$, but it is continuous for $y \neq 0$.

(c) Pushing further the use of Kolmogorov's criterion, it can be shown that Brownian local times $L_t^y (B)$ are Hölder continuous of order $(\frac{1}{2} - \varepsilon)$, for any $\varepsilon > 0$, in the variable y, uniformly in $t \leq T$.

We shall denote by $L_t^y (Y)$ the modification which is continuous in t and right-continuous with left limits in y. Then

$$L_t^{y^-} (Y) = \tilde{L}_t^y (Y) \quad a.s. \quad \text{(see formula (2.1.3))}$$

and the following formulae, known as Tanaka's formulae, hold $a.s.$

$$(Y_t - y)^+ = (Y_0 - y)^+ + \int_0^t dY_s \mathbf{1}_{(Y_s > y)} + \frac{1}{2} L_t^y (Y) \qquad (2.2.1)$$

$$(Y_t - y)^- = (Y_0 - y)^- - \int_0^t dY_s \mathbf{1}_{(Y_s \leq y)} + \frac{1}{2} L_t^y (Y) \qquad (2.2.2)$$

$$|Y_t - y| = |Y_0 - y| + \int_0^t dY_s \mathrm{sgn}(Y_s - y) + L_t^y (Y) \qquad (2.2.3)$$

where $\mathrm{sgn}(x) = 1$ if $x > 0$ and $\mathrm{sgn}(x) = -1$ if $x \leq 0$.

It is useful to know the laws of the occupation measures of B up to τ_l and T_0, where $(\tau_l; l \geq 0)$ is the inverse local time, and $T_0 = \inf\{t : B_t = 0\}$. The laws of $(L^x_{\tau_l}; x \in \mathbb{R})$ and $(L^x_{T_0}; x \in \mathbb{R})$ are given by the Ray–Knight theorems:

Theorem 2.2.4 (Ray–Knight theorems).

(1) $(L^x_{\tau_l}; x \geq 0)$ and $(L^{-x}_{\tau_l}; x \geq 0)$ are the squares of two independent Bessel processes of dimension 0, starting at l and ending at 0, (denoted $BESQ_l(0)$).

(2) If $B_0 = 1$, $(L^x_{T_0}; 0 \leq x \leq 1)$ is distributed as the square of a two-dimensional Bessel process starting at 0.

(We denote by $BESQ_x(\delta)$ a squared Bessel process of dimension δ starting from x, and let Q^δ_x be the law associated with this process).

Note that if $\delta > 0$ is an integer, the process $BESQ_x(\delta)$ may be realized as

$$\left(|a + B_u|^2; u \geq 0\right)$$

where $|a|^2 = x$, $a \in \mathbb{R}^\delta$ and $(B_u; u \geq 0)$ is a BM in $\mathbb{R}^\delta$.

In general, the following convolution property holds:

$$Q^{\delta+\delta'}_{x+x'} = Q^\delta_x * Q^{\delta'}_{x'}, \quad \text{for every } \delta, \delta' \geq 0, \text{ and } x, x' \geq 0. \tag{C}$$

Under Q^δ_x, the canonical process solves the equation

$$X_t = x + 2 \int_0^t \sqrt{X_s} d\beta_s + \delta t, \tag{E}$$

where $(\beta_s; s \geq 0)$ is a BM. From this stochastic differential equation (E), which enjoys the pathwise uniqueness property, property (C) follows easily.

We shall not give the full proof of the Ray–Knight theorems (see, e.g. Jeulin [8]), but only explain why the BESQ processes come in naturally. Indeed, one has:

Proposition 2.2.5. *For any Brownian stopping time S, the Brownian local times $(L^x_S)_{x \in \mathbb{R}}$ satisfy:*

$$P-\lim_{n \to \infty} \sum_{\sigma_n} (L^{x_{i+1}}_S - L^{x_i}_S)^2 = 4 \int_a^b L^x_S dx \tag{2.2.4}$$

where (σ_n) is a sequence of subdivisions of $[a, b]$, whose mesh goes to 0, as $n \to \infty$.

Let us admit Proposition 2.2.5 for a moment. Then, assuming that $(L^x_S)_{a \leq x \leq b}$ is a semimartingale, it can be written as:

$$L^x_S = L^a_S + 2 \int_a^x \sqrt{L^y_S} \, d\beta_y + V_x - V_a,$$

where (β_y) is a Brownian motion, and (V_x) is a process with bounded variation. Note that the diffusion coefficient $2\sqrt{z}$ is precisely that of a BESQ.

Proof of Proposition 2.2.5. It is easily shown that the LHS of (2.2.4) has the same P-limit as $n \to \infty$ as

$$4 \sum_{\sigma_n} \left(\int_0^S dB_u \mathbf{1}_{(x_i < B_u < x_{i+1})} \right)^2$$

thanks to Tanaka's formula (2.2.1). Then, using Itô's formula:

$$Y_S^2 = Y_0^2 + 2 \int_0^S Y_u dY_u + \langle Y \rangle_S,$$

with $Y_t = \int_0^t dB_u \mathbf{1}_{(x_i < B_u < x_{i+1})}$, we prove that the only contribution shall be that of the brackets, namely:

$$4 \sum_{\sigma_n} \int_0^S du\, \mathbf{1}_{(x_i < B_u < x_{i+1})} = 4 \int_a^b L_S^x dx.$$

□

2.3 Occupation Times Are Local Times

We have the following theorem:

Theorem 2.3.1.

(a) $L_t^y = \lim_{\epsilon \to 0} \frac{1}{\epsilon} \int_0^t d\langle Y \rangle_s \mathbf{1}_{(y < Y_s < y+\epsilon)}$

(b) $L_t^{y^-} = \lim_{\epsilon \to 0+} \frac{1}{\epsilon} \int_0^t d\langle Y \rangle_s \mathbf{1}_{(y-\epsilon < Y_s < y)}$

   In particular, $(L_t^y, t \geq 0)$ and $(L_t^{y^-}, t \geq 0)$ are increasing processes.

(c) a.s. the support of the random measure $d_t L_t^y$ is contained in $\{t : Y_t = y\}$.

Proof. (a) and similarly (b) follow from:

$$\frac{1}{\epsilon} \int_0^t d\langle Y \rangle_s \mathbf{1}_{(y < Y_s < y+\epsilon)} = \frac{1}{\epsilon} \int_y^{y+\epsilon} d\zeta L_t^\zeta(Y) \xrightarrow[\epsilon \to 0]{} L_t^y(Y).$$

(c) We use Itô's formula and Tanaka's formula (2.2.3) to write

$$(Y_t - y)^2 = (Y_0 - y)^2 + 2 \int_0^t (Y_s - y)dY_s + \langle Y \rangle_t$$

and, also: $|Y_t - y|^2 = |Y_0 - y|^2 + 2 \int_0^t |Y_s - y| d|Y_s - y| + \langle Y \rangle_t$

$$|Y_t - y| = |Y_0 - y| + 2 \int_0^t \text{sgn}(Y_s - y)dY_s + L_t^y$$

whence

$$(Y_t - y)^2 - (Y_0 - y)^2 = 2 \int_0^t |Y_s - y|\big(\mathrm{sgn}(Y_s - y)dY_s + dL_s^y\big) + \langle Y \rangle_t$$

and then

$$\int_0^t |Y_s - y|\mathrm{sgn}(Y_s - y)dY_s + \int_0^t |Y_s - y|dL_s^y = \int_0^t (Y_s - y)dY_s.$$

Since $x = |x|\mathrm{sgn}(x)$, this implies that:

$$\int_0^t |Y_s - y|dL_s^y = 0.$$

□

2.4   Local Times and the Balayage Formula

Dual Predictable Projection. Given a filtration $(\mathcal{F}_t; t \geq 0)$ and a random time Σ (not necessarily a stopping time), there exists a unique increasing $\mathcal{F}_t$-predictable process $(A_t; t \geq 0)$ such that

$$E[h_\Sigma] = E\Big[\int_0^\infty h_u dA_u\Big]$$

for every predictable process $h \geq 0$.

In the Strasbourg terminology, (A_t) is the dual predictable projection of the process $\mathbf{1}_{(\Sigma \leq t)}$. Here, we shall simply say that (A_t) is the (predictable) increasing process associated to Σ.

The Balayage Formula. Let Y be a continuous semimartingale and define

$$g_t \equiv \sup\{s \leq t : Y_s = 0\}$$

then

$$h_{g_t} Y_t = h_0 Y_0 + \int_0^t h_{g_s} dY_s$$

for every predictable, locally bounded process h.

Proof. By the Monotone Class theorem, it is enough to show this formula for processes of the form $h_u = \mathbf{1}_{[0,T]}(u)$, where T is a stopping time. In this case,

$$h_{g_t} = \mathbf{1}_{(g_t \leq T)} = \mathbf{1}_{(t \leq d_T)},$$

where $d_T = \inf\{s \geq T : Y_s = 0\}$. Hence,

$$h_{g_t} Y_t = \mathbf{1}_{(t \leq d_T)} Y_t = Y_{t \wedge d_T} = \int_0^t \mathbf{1}_{(s \leq d_T)} dY_s = \int_0^t h_{g_s} dY_s.$$

$\square$

Applications.

(a) Let $Y_t = B_t$, then from the balayage formula we obtain that

$$h_{g_t} B_t = \int_0^t h_{g_s} dB_s$$

is a local martingale.

(b) Let $Y_t = |B_t|$. Tanaka's formula (2.2.3) gives

$$|B_t| = \int_0^t \text{sgn}(B_s) dB_s + L_t,$$

where L_t denotes the local time of $(B_t; t \geq 0)$ at $y = 0$. By an application of the balayage formula, we obtain:

$$h_{g_t} |B_t| = \int_0^t h_{g_s} \text{sgn}(B_s) dB_s + \int_0^t h_s dL_s$$

where we have used the fact that dL_s a.s, $h_{g_s} = h_s$. Consequently, replacing, if necessary, h by $|h|$, we see that the process $\int_0^t |h_s| dL_s$ is the local time at 0 of $(h_{g_t} B_t; t \geq 0)$.

Let now T be a stopping time such that $(B_{t \wedge T}; t \geq 0)$ is uniformly integrable, and satisfies: $P(B_T = 0) = 0$. We restrict ourselves to predictable and bounded processes h. Then, it follows from the previous formula that:

$$E[h_{g_T} |B_T|] = E\left[\int_0^T h_s dL_s \right]. \tag{2.4.1}$$

As an example, consider: $T = \tilde{T}_a \equiv \inf\{t : |B_t| = a\}$; we have:

$$E[h_{g_{\tilde{T}_a}}] = \frac{1}{a} E\left[\int_0^{\tilde{T}_a} h_s dL_s \right],$$

whence we conclude that the predictable increasing process $(A_t; t \geq 0)$ associated to $\Sigma = g_{\tilde{T}_a}$ is given by:

$$A_t = \frac{1}{a} L_{t \wedge \tilde{T}_a}.$$

In the general case, applying (2.4.1) to the variable $\xi_{g_T} = E[\,|B_T|\,|\mathcal{F}_{g_T}]$, where $(\xi_u; u \geq 0)$ denotes some predictable process (note that $P(\xi_{g_T} = 0) = 0$, as a consequence of $P(B_T = 0) = 0$) and letting $h' = h\xi$, we obtain:

$$E[h'_{g_T}] = E\Big[\int_0^T \frac{h'_s}{\xi_s} dL_s\Big]$$

from which we deduce that the increasing predictable process associated to $\Sigma = g_T$ is:

$$A_t = \int_0^{t \wedge T} \frac{dL_s}{\xi_s}.$$

In this generality, we have the following interesting lemma:

Lemma 2.4.1 (Azéma). A_T is an exponential variable with mean 1.

Proof. Since $A_T = A_{g_T}$, we have for every $\lambda \geq 0$,

$$E[\exp(-\lambda A_T)] = E\Big[\int_0^T \exp(-\lambda A_s) dA_s\Big],$$

as a consequence of the definition of A. Thus, we obtain:

$$E[\exp(-\lambda A_T)] = E\Big[\frac{1 - \exp(-\lambda A_T)}{\lambda}\Big],$$

or equivalently:

$$E[\exp(-\lambda A_T)] = \frac{1}{1 + \lambda},$$

the desired result follows immediately.   □

Exercise 2.4.2. Deduce the result in Sect. 1.8 from Azéma's lemma 2.4.1, and vice-versa.

Hint: If $g = \sup\{t < T_0 : M_t = S_t\}$, then $P(g > t|\mathcal{F}_t) = M_t/S_t$, $t \leq T_0$; Hence: $A_t = \log S_t$.

From Lemma 2.4.1, we deduce the law of the local time $L_{\tilde{T}_a}$ introduced above.

Corollary 2.4.3. $L_{\tilde{T}_a}$ is an exponential variable, with mean a.

Important Remark. In general, finding ξ may necessitate some work, but in some cases, e.g.: $T = \inf\{t : |B_t| = \rho_t\}$, for a continuous adapted process (ρ_t) such that $\rho_t \equiv \rho_{g_t}$, no extra computation is needed, since: $|B_T| = \rho_{g_T}$ is $\mathcal{F}_{g_T}$ measurable, so that we can take: $\xi_u = \rho_u$; hence, in this case: $A_t = \int_0^{t \wedge T} \frac{dL_s}{\rho_s}$.

2.5   Some Simple Martingales

(a) Let h be bounded and predictable. As we have already seen, the local time at 0 of the martingale $M = (h_{g_t} B_t; t \geq 0)$ is $\int_0^t |h_s| dL_s$. Now, let $h_u \equiv f(L_u) \geq 0$ and let F be the primitive of f with $F(0) = 0$, then

$$F(L_t) = \int_0^t f(L_u) dL_u = \int_0^t h_u dL_u,$$

and also

$$h_{g_t} |B_t| - \int_0^t h_u dL_u$$

is a martingale, and thus $(F(L_t) - f(L_t)|B_t|; t \geq 0)$ is a martingale.

Remark 2.5.1. This result can be extended by linearity when F is not necessarily positive. It can be shown directly using Itô's formula

(b) As a variant of the preceding construction, we consider

$$Y_t = S_t - B_t, \quad \text{and} \quad \gamma_t \equiv \sup\{s < t : B_s = S_s\} = \inf\{s < t : B_s = S_t\}.$$

Letting $f(S_s) = h_s$ and observing that $S_{\gamma_t} = S_t$, from the balayage formula, we get that

$$\left(F(S_t) - (S_t - B_t) f(S_t); t \geq 0 \right)$$

is a martingale, since

$$f(S_t)(S_t - B_t) = f(S_{\gamma_t})(S_t - B_t) = \int_0^t f(S_s) d(S_s - B_s).$$

From these martingales, we can deduce some particular cases of Doob's L^p inequality.

Let M be a continuous martingale and $S_t = \sup_{s \leq t} M_s$. Choose $f(s) = s^{p-1}$, $p > 1$, then by (2.1.2), after localization if necessary,

$$\left(S_t^p - p S_t^{p-1}(S_t - M_t); t \geq 0 \right)$$

is a martingale, and then

$$(p-1) E[S_t^p] = p E[S_t^{p-1} M_t].$$

Doob's inequality now follows from Hölder's inequality and the optional sampling theorem. (It would also be possible to obtain the "full" Doob's inequalities for submartingales using similar arguments.)

2.6   The Existence of Principal Values Related to Brownian Local Times

Let $\alpha \in \mathbb{R}$. We wish to define the process $(H_t^{(\alpha)}; t \geq 0)$

$$H_t^{(\alpha)} \equiv \int_0^t \frac{ds}{B_s^\alpha}$$

where $B_s^\alpha = \operatorname{sgn}(B_s)|B_s|^\alpha$, in a rigorous manner.

Proposition 2.6.1. *For every $\alpha < \frac{3}{2}$,*

$$\lim_{\epsilon \to 0} \int_0^t \frac{ds}{B_s^\alpha} \mathbf{1}_{(|B_s| \geq \epsilon)} \quad \text{exists.}$$

In particular

$$\lim_{\epsilon \to 0} \int_0^t \frac{ds}{B_s} \mathbf{1}_{(|B_s| \geq \epsilon)} \quad \text{exists.}$$

Proof. The occupation density formula implies:

$$\int_0^t \frac{ds}{B_s^\alpha} \mathbf{1}_{(|B_s| \geq \epsilon)} = \int_{\{|a| \geq \epsilon\}} \frac{da}{a^\alpha} L_t^a = \int_\epsilon^\infty \frac{da}{a^\alpha} (L_t^a - L_t^{-a}).$$

Choose a $(\frac{1}{2} - \eta)$-Hölder continuous modification of the local time (from Example 2.2.3(c)) and, in this case, the convergence of the previous integral follows from Cauchy's criterion, since

$$\int_{\epsilon_1}^{\epsilon_2} \frac{da}{|a|^\alpha} |L_t^a - L_t^{-a}| \leq C_{t,\omega} \int_{\epsilon_1}^{\epsilon_2} \frac{da}{a^{\alpha - \frac{1}{2} + \eta}} \longrightarrow 0, \quad \text{as } \epsilon_1, \epsilon_2 \to 0$$

as long as $\alpha < \frac{3}{2}$, by picking η such that $\alpha - \frac{1}{2} + \eta < 1$.   □

Remark 2.6.2. Prove that

$$\int_0^t \frac{ds}{|B_s|^\alpha} = \int_{-\infty}^\infty \frac{da}{|a|^\alpha} L_t^a < \infty \iff \alpha < 1.$$

On the other hand, if $\alpha > \frac{3}{2}$, then

$$\int_\epsilon^\infty \frac{da}{a^\alpha}(L_t^a - L_t^{-a}) = \epsilon^{1-\alpha}\int_1^\infty \frac{dx}{x^\alpha}(L_t^{\epsilon x} - L_t^{-\epsilon x})$$

$$= \epsilon^{\frac{3}{2}-\alpha}\int_1^\infty \frac{dx}{x^\alpha}\Big(\frac{1}{\sqrt{\epsilon}}(L_t^{\epsilon x} - L_t^{-\epsilon x})\Big).$$

It can be shown that

$$\frac{1}{\sqrt{\epsilon}}(L_t^{\epsilon x} - L_t^{-\epsilon x}) \xrightarrow{\text{(law)}} \mathbf{B}_{(x,L_t^0)} \quad \text{as } (\epsilon \to 0)$$

where $\mathbf{B}_{(x,y)}$ denotes a Brownian sheet independent of L_t^0, which yields a limit theorem for the above integrals.

2.7 Some Extensions of Itô's Formula

(a) Consider Tanaka's formula for a continuous semimartingale Y,

$$(Y_t - y)^+ = (Y_0 - y)^+ + \int_0^t dY_s \mathbf{1}_{(Y_s > y)} + \frac{1}{2}L_t^y.$$

We integrate this formula with respect to a Radon measure $\mu(dy)$, which, for the moment, we assume to be bounded. Then

$$\int \mu(dy)\big[(Y_t - y)^+ - (Y_0 - y)^+\big] = \int_0^t dY_s \mu(]-\infty, Y_s[) + \frac{1}{2}\int \mu(dy)L_t^y.$$

Using a localization argument, the previous formula can be generalized for any Radon measure $\mu(dx) = F''(dx)$; thus:

$$F(Y_t) = F(Y_0) + \int_0^t dY_s F_l'(Y_s) + \frac{1}{2}\int F''(dy)L_t^y,$$

where F_l' denotes the left-derivative of F.

Example 2.7.1. If F is C^2 except at a finite number of points $x_1, x_2, \ldots, x_k$ where F' admits right and left limits, and F'' is locally integrable, we have:

$$F(B_t) = F(B_0) + \int_0^t F'(B_s)dB_s + \frac{1}{2}\sum_{i=1}^k \big(F'(x_i+) - F'(x_i-)\big)L_t^{x_i}$$

$$+ \frac{1}{2}\int_0^t ds F''(B_s).$$

As a particular case, we obtain the following formula for $F_k(x) = \text{dist}(x; \{x_1, \ldots, x_k\})$:

$$\text{dist}(B_t; \{x_1, \ldots, x_k\}) = \text{dist}(B_0; \{x_1, \ldots, x_k\}) + B_t^\epsilon + \sum_{i=1}^{k} L_t^{x_i}(B) - \sum_{i=1}^{k-1} L_t^{y_i}(B),$$

where

$$y_i = \frac{x_i + x_{i+1}}{2}, \quad (1 < i \le k - 1),$$

$$B_t^\epsilon = \int_0^t \epsilon(B_s) dB_s, \quad \text{with}$$

$$\epsilon(y) = -\mathbf{1}_{(y < x_1)} + \mathbf{1}_{(x_1 < y < y_1)} - \mathbf{1}_{(y_1 < y < x_2)} + \cdots + \mathbf{1}_{(x_n < y)}$$

(b) Itô's formula can also be extended to BM on $\mathbb{R}^n$, for functions F such that $\triangle F = \mu$ is a Radon measure. The additive functional $A^{(\mu)}$ associated to μ through the Revuz bijection $\mu \leftrightarrow A^{(\mu)}$, (see Brosamler and Meyer), shows up in this extension of Itô's formula:

$$F(B_t) = F(B_0) + \int_0^t \nabla F(B_s) \cdot dB_s + \frac{1}{2} A_t^{(\mu)}.$$

(c) Let us consider another extension of Itô's formula. Let B be a one-dimensional BM and consider

$$F(x) = F(0) + \int_0^x dy f(y)$$

where $f \in L_{\text{loc}}^2(\mathbb{R})$, then

$$F(B_t) = F(B_0) + \int_0^t dB_s f(B_s) + \frac{1}{2} A_t^{(f)}.$$

The process $\{A_t^{(f)}; t \ge 0\}$ has the following properties:

(i) $A_t^{(f)}$ has zero quadratic variation.
(ii) If f is differentiable then

$$A_t^{(f)} = \int dy f'(y) L_t^y = \int_0^t ds f'(B_s) = -\int d_y L_t^y f(y).$$

(iii) In the general case, $A_t^{(f)} = -\int d_y L_t^y f(y)$ since $(L_t^y; y \in \mathbb{R})$ is a semimartingale in the space-parameter, a result obtained by Perkins and Jeulin.

(d) Föllmer–Protter–Shyriaev showed that, if $f \in L^2_{loc}(\mathbb{R})$, then

$$A_t^{(f)} = \langle f(B), B \rangle_t \equiv \lim_{\tau_n} \sum_{\tau_n} [f(B_{t_{i+1}}) - f(B_{t_i})](B_{t_{i+1}} - B_{t_i})$$

where $\{\tau_n\}$ is a sequence of partitions of $[0, t]$ whose diameter goes to zero.

Applications.

(i) Let $F(x) = x \log |x| - x$; then: $f(x) = \log |x|$; hence, in this case

$$A_t^{(f)} = \int_0^t \frac{ds}{B_s},$$

$$F(B_t) = \int_0^t \log |B_s| dB_s + \frac{1}{2} \int_0^t \frac{ds}{B_s}.$$

(ii) Let $f(x) = \mathbf{1}_{(x>0)}$. We then obtain the following approximation of the local time:

$$A_t^{(f)} = \lim_{\tau_n} \sum_{\tau_n} \left(\mathbf{1}_{(B_{t_{i+1}}>0)} - \mathbf{1}_{(B_{t_i}>0)} \right) (B_{t_{i+1}} - B_{t_i}).$$

Exercise 2.7.2.

(a) With the same hypothesis as in Exercise 1.8.2, prove that

$$P\left(g_\infty^{(a)} \le t\right) = \frac{1}{2} E\left[L_t^a(M)\right]. \tag{2.7.1}$$

(b) We assume both following hypotheses:

(i) $d\langle M \rangle_s = \sigma_s^2 ds$, such that $E[\sigma_s^2 | M_s = x]$ may be chosen jointly continuous in (s, x);

(ii)   for every $s \ge 0$, the law of M_s admits a density $m_s(x)$, which also may be chosen jointly continuous in (s, x).

Prove that under these hypotheses, for every $a \ge 0$:

$$E[L_t^a] = \int_0^t ds \, E[\sigma_s^2 | M_s = a] m_s(a). \tag{2.7.2}$$

(c) Deduce the equality:

$$P\left(g_\infty^{(a)} \in dt\right) = \frac{dt}{2} E[\sigma_t^2 | M_t = a] m_t(a). \tag{2.7.3}$$

(d) Apply the preceding results to show the identity:

$$E\left[\left|\exp\left(B_t - \frac{t}{2}\right) - 1\right|\right] = 2P(4B_1^2 \leq t). \qquad (2.7.4)$$

Note that since $\left(\exp\left(B_t - \frac{t}{2}\right) - 1; t \geq 0\right)$ is a true martingale, $E\left[1 - \exp\left(B_t - \frac{t}{2}\right)\right] = 0$; hence:

$$E\left[\left|\exp\left(B_t - \frac{t}{2}\right) - 1\right|\right] = 2E\left[\left(1 - \exp\left(B_t - \frac{t}{2}\right)\right)^+\right].$$

Hence, we need to prove:

$$E\left[\left(1 - \exp\left(B_t - \frac{t}{2}\right)\right)^+\right] = P(4B_1^2 \leq t), \qquad (2.7.5)$$

that is,

$$P\left(g^{(1)}\left(\exp\left(B_u - \frac{u}{2}\right); u \geq 0\right) \leq t\right) = P(4B_1^2 \leq t).$$

Note that

$$g^{(1)}\left(\exp\left(B_u - \frac{u}{2}\right); u \geq 0\right) \equiv \sup\left\{u : B_u - \frac{u}{2} = 0\right\}$$

$$\equiv \sup\left\{u : \frac{1}{u}B_u = \frac{1}{2}\right\}$$

$$= 1/\inf\left\{t : tB_{1/t} = \frac{1}{2}\right\}$$

$$\overset{(\text{law})}{=} 1/T_{(1/2)}(B)$$

$$\overset{(\text{law})}{=} 4B_1^2.$$

Thus, we have proven (2.7.5).

References

1. P.-A. Meyer, Un cours sur les intégrales stochastiques. Séminaire de Probabilités, X (Seconde partie: Théorie des intégrales stochastiques, Univ. Strasbourg, Strasbourg, année universitaire 1974/1975). *Lecture Notes in Math.*, vol. 511. (Springer, Berlin, 1976), pp. 245–400
2. M. Yor, Sur la continuité des temps locaux associés à certaines semimartingales. In: *Temps Locaux*, Astérisque, 52–53, 23–36, 1978
3. J. Azéma, M. Yor, En guise d'introduction. In: *Temps Locaux*, Astérisque, 52–53, 3–16, 1978

4. J. Azéma, M. Yor, Une solution simple au problème de Skorokhod. Séminaire de Probabilités, XIII (Univ. Strasbourg, Strasbourg, 1977/78). *Lecture Notes in Math.*, vol. 721. (Springer, Berlin, 1979), pp. 90–115

5. T. Yamada, On the fractional derivative of Brownian local times. J. Math. Kyoto Univ. **25**(1), 49–58 (1985)

6. G.A. Brosamler, Quadratic variation of potentials and harmonic functions. Trans. Am. Math. Soc. **149**, 243–257 (1970)

7. H. Föllmer, P. Protter, A.N. Shiryayev, Quadratic covariation and an extension of Itô's formula. Bernoulli **1**(1–2), 149–169 (1995)

8. T. Jeulin, Ray-Knight's theorem on Brownian local times and Tanaka's formula. Seminar on stochastic processes, 1983 (Gainesville, Fla., 1983). *Progr. Probab. Statist.*, vol. 7. (Birkhäuser, Boston, 1984), pp. 131–142

9. P.-A. Meyer, La formule d'Itô pour le mouvement brownien d'après G. Brosamler. Séminaire de Probabilités, XII (Univ. Strasbourg, Strasbourg, 1976/1977). *Lecture Notes in Math.*, vol. 629. (Springer, Berlin, 1978), pp. 763–769

10. E. Perkins, Local time is a semimartingale. Z. Wahrsch. Verw. Gebiete, **60**(1), 79–117 (1982)

11. M. Yor, Sur la transformée de Hilbert des temps locaux browniens, et une extension de la formule d'Itô. Seminar on Probability, XVI. *Lecture Notes in Math.*, vol. 920. (Springer, Berlin, 1982), pp. 238–247

Chapter 3
Lévy's Representation of Reflecting BM and Pitman's Representation of BES(3)

We show how the reflected BM and BES(3) process may be represented with the help of the one-sided supremum of Brownian motion.

3.1 Lévy's Identity in Law: The Local Time as a Supremum Process

The process $(|B_t|; t \geq 0)$ is called reflected Brownian motion.

Theorem 3.1.1 (Lévy). *The following identity in law holds*

$$(|B_t|, L_t; t \geq 0) \overset{(law)}{=} (S_t - B_t, S_t; t \geq 0).$$

Proof. Compare the two decompositions:

$$|B_t| = \int_0^t \operatorname{sgn}(B_s)dB_s + L_t,$$

$$S_t - B_t = (-B_t) + S_t.$$

Then, since $\int_0^t \operatorname{sgn}(B_s)dB_s$ is a BM, the theorem will be obtained if we show that

$$L_t = \sup_{s \leq t}\left(-\int_0^s \operatorname{sgn}(B_u)dB_u\right).$$

This is the consequence of the following: ☐

Lemma 3.1.2 (Skorokhod). *Let $(y(u); u \geq 0)$ be a given continuous function with $y(0) = 0$, and consider as unknown a pair of continuous functions $(x(u), \ell_u \equiv \ell(u); u \geq 0)$ such that:*

J.-Y. Yen and M. Yor, *Local Times and Excursion Theory for Brownian Motion*, Lecture Notes in Mathematics 2088, DOI 10.1007/978-3-319-01270-4_3, © Springer International Publishing Switzerland 2013

(i) $\forall u \geq 0, x(u) = -y(u) + \ell_u$

(ii) $\forall u, x(u) \geq 0$

(iii) $(\ell_u, u \geq 0)$ is increasing and the support of $d\ell_u$ is contained in the set of zeros of x.

Then, there exist a unique (x^*, ℓ^*) which satisfies (i), (ii), and (iii). This solution is given by:

$$\ell_u^* = \sup_{0 \leq s \leq u} y(s) \ \text{ and } \ x^*(u) = -y(u) + \ell_u^*.$$

Proof. It is easy to see that the pair (x^*, ℓ^*) verifies (i), (ii), and (iii). To establish uniqueness, consider two solutions, (x_1, ℓ_1) and (x_2, ℓ_2). Then

$$0 \leq \big(x_1(u) - x_2(u)\big)^2 = 2 \int_0^u \big(x_1(v) - x_2(v)\big)\big(d\ell_1(v) - d\ell_2(v)\big)$$

$$= -2 \int_0^u x_1(v) d\ell_2(v) - 2 \int_0^u x_2(v) d\ell_1(v) \leq 0,$$

which implies $x_1 = x_2$ and $\ell_1 = \ell_2$.   □

Remark 3.1.3. In practice, Lévy's theorem 3.1.1 is used to "transfer" results bearing on S to L and vice versa.

3.2 A Solution to Skorokhod's Embedding Problem

Problem. Given a probability measure $\mu(dx)$ in $\mathbb{R}$, find a stopping time T of $(B_t; t \geq 0)$ such that B_T has law μ.

Stated without further constraints, the problem has many (non interesting) solutions; for example, if $\mu = \delta_a$, then $T_a^{(1)} = \inf\{t : B_t = a\}$, $T_a^{(2)} = \inf\{t \geq d_{T_a} : B_t = a\}$, ...,etc., are solutions to the problem. However, in a gambler's perspective, say, it is desirable that T be as small as possible. Hence, one usually asks the following:

Additional condition: $(B_{T \wedge t}; t \geq 0)$ is required to be uniformly integrable.
This condition implies that

$$\int |x| d\mu(x) < \infty \ \text{ and } \ \int x d\mu(x) = 0.$$

Even with this additional condition, Skorokhod's problem does not have in general a unique solution. A thorough survey of the different solutions to Skorokhod's problem is presented by J. Obłój [1].

Fig. 3.1 B_{T_μ} is distributed
as μ

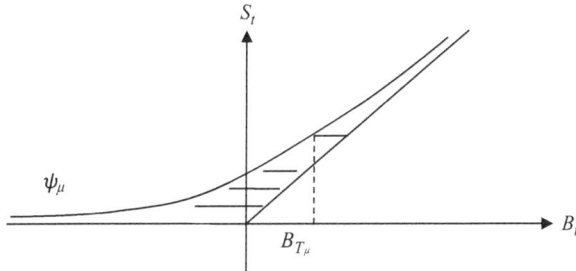

Several authors, Dubins, Root, Rost …, have constructed solutions by iterative and approximation methods. Dubins, Daniels, Siegmund and Lerche give various solutions, some of which of the type

$$T = \inf\{t : B_t = f(t)\}$$

using the Heat Equation.

The proof we give is based on the balayage formula and, in the end, relies simply on solving a first order differential equation.

Theorem 3.2.1. *If μ has a first moment and is centered, then $T_\mu \equiv \inf\{t : S_t \geq \psi_\mu(B_t)\}$ where*

$$\psi_\mu(x) = \frac{1}{\mu[x,\infty)} \int_{[x,\infty)} t\, d\mu(t)$$

is a solution to Skorokhod's problem, i.e., B_{T_μ} has law μ, and $(B_{T_\mu \wedge t}; t \geq 0)$ is uniformly integrable.

Example 3.2.2.

(a) If $T_\mu = \inf\{t : 2S_t - B_t = a\}$, then

$$\psi_\mu(x) = \begin{cases} 0, & \text{for } x < -a \\ \frac{x+a}{2}, & \text{for } -a \leq x \leq a \\ x, & \text{for } x > a \end{cases}$$

(b) If $\mu(dx) = \frac{e^{-x^2/2}}{\sqrt{2\pi}} dx$, then $\psi_\mu(x) = -\frac{\phi'(x)}{\phi(x)}$ where $\phi(x) = \int_x^\infty \frac{1}{\sqrt{2\pi}} e^{-u^2/2} du$.

Sketch of the proof of Theorem 3.2.1. Instead of proving in detail that B_{T_μ} has law μ, we shall prove that if $T_\psi \equiv \inf\{t : S_t \geq \psi(B_t)\}$, for some increasing function ψ, and if B_{T_ψ} has distribution μ, then $\psi = \psi_\mu$ (Fig. 3.1).

By Sect. 2.5, for g C^1 with compact support, $(g(S_t) - g'(S_t)(S_t - B_t); t \geq 0)$ is a martingale, thus denoting simply T_ψ by T, we have:

$$g(0) = E\big[g(S_T) - g'(S_T)(S_T - B_T)\big].$$

If $\psi^{-1} = \varphi$, $B_T = \varphi(S_T)$ and thus

$$g(0) = E\big[g(S_T) - g'(S_T)(S_T - \varphi(S_T))\big].$$

Define $\overline{\gamma}(x) \equiv P(S_T \geq x)$; from the previous equation, we get

$$-\frac{\overline{\gamma}(x)dx}{x - \varphi(x)} = d\overline{\gamma}(x);$$

let now

$$\overline{\mu}(x) = \mu([x, \infty)) = P(B_T \geq x).$$

Then $\overline{\mu}(x) = P\big(\psi(B_T) \geq \psi(x)\big) = P\big(S_T \geq \psi(x)\big) = \overline{\gamma}(\psi(x))$, whence,

$$-d\mu(x) = d\overline{\mu}(x) = d\big(\overline{\gamma}(\psi(x))\big) = -\frac{\overline{\gamma}(\psi(x))d\psi(x)}{\psi(x) - x} = -\frac{\overline{\mu}(x)d\psi(x)}{\psi(x) - x},$$

so

$$\frac{\overline{\mu}(x)d\psi(x)}{\psi(x) - x} = d\mu(x) = -d\overline{\mu}(x).$$

Finally,

$$\psi(x)d\overline{\mu}(x) + \overline{\mu}(x)d\psi(x) = -xd\mu(x),$$

that is, $d(\psi\overline{\mu})(x) = -xd\mu(x)$ and we get $\psi(x)\overline{\mu}(x) = \int_{[x,\infty)} td\mu(t)$, i.e.: $\psi = \psi_\mu$.
The theorem is proved reasoning backwards.

Remark 3.2.3. It is also possible to calculate the law of T_μ and of functionals of the type $\int_0^{T_\mu} dsf(B_s)$.

Example 3.2.4 (We take up Example 3.2.2(a)). Let $\theta_a \equiv \inf\{t : 2S_t - B_t = a\}$. Then θ_a is independent of S_{θ_a}, and S_{θ_a} is uniform on $[0, a]$; indeed, $S_{\theta_a} - B_{\theta_a}$ is uniform on $[0, a]$, and independent from θ_a (we shall see this later on, in Sect. 3.3). Since

$$\cosh\big(\lambda(S_t - B_t)\big)\exp\Big(-\frac{\lambda^2}{2}t\Big)$$

is a martingale, and

$$S_{\theta_a} + (S_{\theta_a} - B_{\theta_a}) = a;$$

one has:

$$E\Big[\exp\Big(-\frac{\lambda^2\theta_a}{2}\Big)\Big] = \frac{1}{\frac{1}{a}\int_0^a dx\cosh(\lambda x)} = \frac{\lambda a}{\sinh(\lambda a)}.$$

Fig. 3.2 Graph of
Example 3.2.2(a)

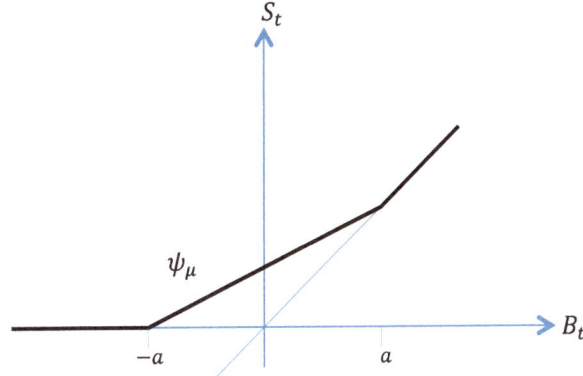

On the other hand,

$$E\left[\exp\left(-\frac{\lambda^2}{2}T_a^{(3)}\right)\right] = \frac{\lambda a}{\sinh(\lambda a)},$$

where $T_a^{(3)} \equiv \inf\{t : |B_t^{(3)}| = a\}$ and $(B_t^{(3)}; t \geq 0)$ denotes a three-dimensional BM. Thus, we have shown that $\theta_a \overset{(law)}{=} T_a^{(3)}$ (see Sect. 1.6; Fig. 3.2).

The representation of the BES(3) process discussed in the following section gives a good explanation of the last identity.

3.3 Pitman's Representation of BES(3)

Theorem 3.3.1 (Pitman). *The following identity in law holds*

$$(2S_t - B_t, S_t; t \geq 0) \overset{(law)}{=} (R_t, J_t; t \geq 0)$$

where $(R_t; t \geq 0)$ is a BES(3) process and $J_t \equiv \inf_{s \geq t} R_s$.

Proof. Lévy's identity in law (Theorem 3.1.1) implies that the following two statements are equivalent:

$$(2S_t - B_t; t \geq 0) \overset{(law)}{=} (R_t; t \geq 0)$$

and

$$(|B_t| + L_t; t \geq 0) \overset{(law)}{=} (R_t; t \geq 0).$$

By a change of scale, we only need to prove that

$$(|B_t| + L_t; t \le \tau_1) \stackrel{(\text{law})}{=} (R_t; t \le \gamma^1) \tag{P}$$

where $\tau_1 \equiv \inf\{t : L_t \ge 1\}$ and $\gamma^1 \equiv \sup\{t : R_t = 1\}$.

Since the Brownian bridge is stable under time reversal, we get, from Sect. 4.3, Example 4.3.1(b), that:

$$(B_{\tau_1-t}; t \le \tau_1) \stackrel{(\text{law})}{=} (B_t; t \le \tau_1).[1]$$

This holds jointly with

$$(1 - L_{\tau_1-t}; t \le \tau_1) \stackrel{(\text{law})}{=} (L_t; t \le \tau_1).$$

Then, (P) is equivalent to:

$$\left(|B_{\tau_1-t}| + (1 - L_{\tau_1-t}); t \le \tau_1\right) \stackrel{(\text{law})}{=} (R_t; t \le \gamma^1).$$

By Lévy's identity, this is equivalent to

$$\left(1 - B_{T_1-t}; t \le T_1\right) \stackrel{(\text{law})}{=} (R_t; t \le \gamma^1)$$

which follows from Williams' time reversal (see Sect. 1.6).   □

Corollary 3.3.2. *Let $\rho_t = 2S_t - B_t$, $\mathcal{R}_t = \sigma\{\rho_s; s \le t\}$, and let T be a $\mathcal{R}_t$-stopping time. Then, conditionally on $\mathcal{R}_T$, the r.v. S_T (and, consequently, the r.v. $S_T - B_T$) is uniformly distributed on $[0, \rho_T]$.*

Proof. Using Pitman's theorem 3.3.1, the statement of the corollary is equivalent to: If $(R_s^a; s \ge 0)$ is a $\text{BES}_a(3)$ process, $\inf_{s \ge 0} R_s^{(a)}$ is uniform on $[0, a]$, which follows from Lemma 1.8.1 applied to the local martingale $M = 1/R$.   □

---

[1] It is even more direct to see this identity as a consequence of the switching identity:

$$E[F(B_u; u \le \tau_1)|\tau_1 = t] = E[F(B_u; u \le t)|B_t = 0, L_t = 1].$$

3.4 A Relation Between (The Above Solution to) Skorokhod's Problem and the Balayage Formula

Keeping with the notation in Sect. 3.2, we consider again:
$T_\mu = \inf\{t : S_t \geq \psi_\mu(B_t)\}$ and we introduce: $\gamma_\mu = \sup\{t \leq T_\mu : S_t - B_t = 0\}$;
we want to calculate the increasing process associated to γ_μ. The balayage formula applied to $Y = S - B$ yields:

$$E[h_{\gamma_\mu}(S_{T_\mu} - B_{T_\mu})] = E\Big[\int_0^{T_\mu} h_v dS_v\Big]$$

where h is any positive predictable bounded process.

Let $\varphi = \psi^{-1}$; then: $S_{T_\mu} - B_{T_\mu} = \psi_\mu(B_{T_\mu}) - B_{T_\mu} = S_{T_\mu} - \varphi(S_{T_\mu})$, hence

$$E[h_{\gamma_\mu}(S_{\gamma_\mu} - \varphi(S_{\gamma_\mu}))] = E\Big[\int_0^{T_\mu} h_v dS_v\Big].$$

Denoting $k = h(S - \varphi(S))$, we get:

$$E[k_{\gamma_\mu}] = E\Big[\int_0^{T_\mu} \frac{k_v}{S_v - \varphi(S_v)} dS_v\Big],$$

whence

$$A_t = \int_0^{T_\mu \wedge t} \frac{dS_v}{S_v - \varphi(S_v)} \equiv \theta(S_{t \wedge T_\mu}),$$

where θ is defined by:

$$\theta(x) = \int_0^x \frac{dy}{y - \varphi(y)}.$$

Hence, $\theta(S_{t \wedge T_\mu})$ is the increasing process associated to γ_μ.

Exercise 3.4.1. Define $T'_\varphi = \inf\{t : |B_t| \geq \varphi(L_t)\}$ for a certain function φ. Find the increasing predictable process associated to $g_{T'_\varphi} \equiv \sup\{t < T'_\varphi : B_t = 0\}$.

Exercise 3.4.2. Deduce from the above computation of (A_t) and from Azéma's lemma 2.4.1 (i.e. A_{T_μ} is exponentially distributed with mean 1) that T_μ solves Skorokhod's problem.

Hint: We know from Azéma's lemma: $P(S_{T_\mu} \geq x) = \exp(-\theta(x))$ which implies $B_{T_\mu} \sim \mu$. (See the proof of Theorem 3.2.1.)

3.5   An Extension of Pitman's Theorem to Brownian Motion with Drift

Let $\mu \in \mathbb{R}$ and denote

$$B_t^{(\mu)} = B_t + \mu t, \qquad S_t^{(\mu)} = \sup_{s \le t} B_s^{(\mu)}$$

and $L_t^{(\mu)}$ the local time of $B_t^{(\mu)}$ at zero.

Theorem 3.5.1. *The processes* $(|B_t^{(\mu)}| + L_t^{(\mu)}; t \ge 0)$ *and* $(2S_t^{(\mu)} - B_t^{(\mu)}; t \ge 0)$ *have the same distribution as the diffusion process with infinitesimal generator*

$$\frac{1}{2}\frac{d^2}{dx^2} + \mu \coth(\mu x)\frac{d}{dx}.$$

Proof. Let F be a $\mathcal{R}_t$-measurable functional, where $\mathcal{R}_t = \sigma\{\rho_s \equiv |B_s| + L_s; s \le t\}$. Then

$$E\big[F(|B_s^{(\mu)}| + L_s^{(\mu)}; s \le t)\big] = E\Big[F(\rho_s; s \le t)\exp\big(\mu B_t - \frac{\mu^2}{2}t\big)\Big].$$

From the symmetry of the law of $(B_t; t \ge 0)$, we deduce:

$$E\big[F(\rho_s; s \le t)e^{\mu B_t}\big] = E\big[F(\rho_s; s \le t)e^{-\mu B_t}\big],$$

and thus

$$E\big[F(\rho_s; s \le t)e^{\mu B_t}\big] = E\big[F(\rho_s; s \le t)\cosh(\mu B_t)\big].$$

Then,

$$E\Big[F(\rho_s; s \le t)\exp\big(\mu B_t - \frac{\mu^2}{2}t\big)\Big] = E\big[F(\rho_s; s \le t)E[\cosh(\mu|B_t|)|\mathcal{R}_t]\big]e^{-\frac{\mu^2}{2}t}.$$

Using Lévy's identity (Theorem 3.1.1) and Pitman's corollary 3.3.2, we obtain:

$$E[\cosh(\mu|B_t|)|\mathcal{R}_t] = \int_0^r \frac{dx\cosh(\mu x)}{r} = \frac{\sinh(\mu r)}{\mu r}, \quad \text{with } r = \rho_t,$$

whence

$$E\Big[F(\rho_s; s \le t)\exp\big(\mu B_t - \frac{\mu^2}{2}t\big)\Big] = E\Big[F(\rho_s; s \le t)\frac{\sinh \mu\rho_t}{\mu\rho_t}e^{-\frac{\mu^2}{2}t}\Big].$$

On the other hand,

$$E\big[F(2S_s^{(\mu)} - B_s^{(\mu)}; s \le t)\big] = E\big[F(\tilde{\rho}_s; s \le t)e^{\mu B_t - \frac{\mu^2}{2}t}\big]$$

where $\tilde{\rho}_t = 2S_t - B_t$.

Pitman's corollary 3.3.2 implies that, conditionally on $\tilde{\mathcal{R}}_t = \sigma\{\tilde{\rho}_s; s \le t\}$,

$$B_t \overset{\text{(law)}}{=} \tilde{\rho}_t U - \tilde{\rho}_t(1 - U) = \tilde{\rho}_t V$$

where U is uniform in $[0, 1]$ and V is uniform in $[-1, 1]$; thus

$$E\big[e^{\mu B_t} | \tilde{\rho}_t = r\big] = \frac{1}{2r} \int_{-1}^{1} dx\, e^{\mu r x} = \frac{\sinh(\mu r)}{\mu r}$$

which yields:

$$E\big[F(\tilde{\rho}_s; s \le t)e^{\mu B_t - \frac{\mu^2}{2}t}\big] = E\big[F(\tilde{\rho}_s; s \le t)\frac{\sinh(\mu \tilde{\rho}_t)}{\mu \tilde{\rho}_t}e^{-\frac{\mu^2}{2}t}\big].$$

The theorem now follows from Girsanov's theorem.   □

At this point, it is natural to ask what happens to Lévy's Theorem 3.1.1 for reflected BM, when B is changed to $B^{(\mu)}$. In fact, $(|B_t^{(\mu)}|; t \ge 0)$ and $(S_t^{(\mu)} - B_t^{(\mu)}; t \ge 0)$ do not have the same law. This can be easily seen since, on one hand, for $\mu > 0$, $B_t^{(\mu)} \to \infty$ and, on the other hand, $(S_t^{(\mu)} - B_t^{(\mu)}; t \to \infty)$ vanishes for some arbitrarily large t's.

More precisely, we have the following relation:

Theorem 3.5.2. *Let $\mu > 0$, then*

$$\left(|B_s^{(\mu)}|; s \le \tau_t^{(\mu)} \big| \tau_t^{(\mu)} < \infty\right) \overset{\text{(law)}}{=} \left(S_s^{(\mu)} - B_s^{(\mu)}; s \le T_t^{(\mu)}\right)$$

where $T_t^{(\mu)} \equiv \inf\{s \ge 0 : B_s^{(\mu)} = t\}$ and $\tau_t^{(\mu)} \equiv \inf\{s \ge 0 : L_s^{(\mu)} > t\}$.

Proof. Let X be the canonical process and $\tau_t \equiv \inf\{s \ge 0 : L_s(X) > t\}$. We have:

$$\mathbf{W}^{(\mu)}\big[F(|X_s|; s \le \tau_t)\mathbf{1}_{(\tau_t < \infty)}\big] = \mathbf{W}\big[F(|X_s|; s \le \tau_t)e^{-\frac{\mu^2}{2}\tau_t}\big]$$

since $X_{\tau_t} \equiv 0$ under $\mathbf{W}$. Taking $F \equiv 1$ in the previous expression, we see that

$$\mathbf{W}^{(\mu)}(\tau_t < \infty) = \mathbf{W}(e^{-\frac{\mu^2}{2}\tau_t}) = \mathbf{W}(e^{-\frac{\mu^2}{2}T_t}) = e^{-\mu t},$$

since under $\mathbf{W}$, $\tau_t \overset{\text{(law)}}{=} T_t$. We get:

$$\mathbf{W}^{(\mu)}\big[F(|X_s|; s \le \tau_t)\big| \tau_t < \infty\big] = \mathbf{W}\big[F(|X_s|; s \le \tau_t)e^{\mu t - \frac{\mu^2}{2}\tau_t}\big].$$

Using again Lévy's identity, the above expectation is equal to

$$\mathbf{W}\big[F(S_s - X_s; s \leq T_t)e^{\mu t - \frac{\mu^2}{2}T_t}\big] = \mathbf{W}^{(\mu)}\big[F(S_s - X_s; s \leq T_t)\big]. \qquad \square$$

Exercise 3.5.3. Let $(B_t^{(\mu)}; t \geq 0)$ be a BM in $\mathbb{R}$, starting from 0 with drift $\mu \in \mathbb{R}$; then $(|B_t^{(\mu)}|; t \geq 0)$ has the same law as the reflecting diffusion

$$Z_t = \hat{B}_t + \mu \int_0^t \tanh(\mu Z_s) ds + L_t(Z),$$

where $\mu = |\mu|$ and $(\hat{B}_t; t \geq 0)$ is a one-dimensional BM.

Hint: Use Girsanov's theorem and the symmetry of BM.

Exercise 3.5.4. Admit that, conditionally on $\mathcal{R}_t$, S_t is uniform in $[0, \rho_t]$ where $\rho_t = 2S_t - B_t$. Prove Pitman's theorem 3.3.1 using only stochastic calculus.

Hint: Develop ρ_t^2 in the Brownian filtration and project all terms so obtained on $\mathcal{R}_t$. Then, check that ρ_t^2 is a BESQ(3).

3.6 Skorokhod's Lemma and the Balayage Formula

Assume that the given data in Skorokhod's lemma (or reflection equation) is a semimartingale $(Y_t; t \geq 0)$.

Then, the solution (X_t, L_t) to Skorokhod's equation is

$$X_t = -Y_t + L_t \quad \text{and} \quad L_t = \sup_{0 \leq s \leq t} Y_s.$$

We then consider Skorokhod's equation with data:

$$Y_t' = \int_0^t h_{\gamma_s} dY_s,$$

where h is a locally bounded previsible, $\mathbb{R}_+$ valued, process, and $\gamma_s \equiv \sup\{u \leq s; X_u = 0\}$. Then it follows from the balayage formula that the (unique) solution (X_t', L_t') to Skorokhod equation is

$$(X_t' \equiv) \ h_{\gamma_t} X_t = -\int_0^t h_{\gamma_s} dY_s + \int_0^t h_s dL_s \ (\equiv -Y_t' + L_t'). \tag{3.6.1}$$

Hence, we have:

$$\int_0^t h_s \, dL_s = \sup_{s \le t} \left(\int_0^s h_{\gamma_u} \, dY_u \right). \tag{3.6.2}$$

This remark applies in particular to $X_t = S_t - B_t$ and $h_t = f(S_t)$, for $f : \mathbb{R}_+ \to \mathbb{R}_+$ and we can consider the equality:

$$f(S_t)(S_t - B_t) = -\int_0^t f(S_s) \, dB_s + F(S_t)$$

where $F(y) = \int_0^x dy f(y)$, as a particular case of (3.6.1); (3.6.2) becomes:

$$F(S_t) = \sup_{s \le t} \left(\int_0^s f(S_u) \, dB_u \right).$$

Below, we shall make some use of the following combination of Skorokhod's lemma and Pitman's theorem.

Proposition 3.6.1. *Let $f : \mathbb{R}_+ \to \mathbb{R}_+$ be measurable, and locally bounded; define:*

$$F(x) = \int_0^x dy f(y).$$

Then,

(i) *the process: $(F(S_t) - f(S_t)(S_t - B_t); t \ge 0)$ is a local martingale;*

(ii) *the process $(\phi_t = f(S_t)(S_t - B_t); t \ge 0)$ is a time-changed reflecting Brownian motion; more precisely,*

$$\phi_t = \beta \left(\int_0^t du \, f^2(S_u) \right); \quad t \ge 0$$

where $(\beta(u); u \ge 0)$ is a reflecting BM, with local time $(l(u); u \ge 0)$; moreover,

$$F(S_t) = l \left(\int_0^t du \, f^2(S_u) \right); \quad t \ge 0.$$

(iii) *the process $(\Phi_t = F(S_t) + f(S_t)(S_t - B_t); t \ge 0)$ is a time changed BES(3), more precisely,*

$$\Phi_t = R \left(\int_0^t ds \, f^2(S_s) \right); \quad t \ge 0, \tag{3.6.3}$$

where $(R(u); u \ge 0)$ is a three-dimensional Bessel process.

Remark 3.6.2.

1. In Chap. 4, we shall develop some variation on this theme with $X'_t = B_t^+$ and $X''_t = B_t^-$, (see Sect. 4.2).
2. In the particular case where $f(x) = x^p$, it would be interesting, in order to understand better the representation (3.6.3), to study the joint law of $(S_t, B_t, \int_0^t ds S_s^{2p})$, which should be facilitated by the Brownian scaling property.

3.7 Seshadri's Remark on the Joint Law of (S_t, B_t)

We start with the following elementary computation:

$$(2S_t - B_t)^2 = (S_t - B_t)^2 + r_t^2 = S_t^2 + \tilde{r}_t^2,$$

where $\tilde{r}_t^2 = (S_t - B_t)^2 + 2S_t(S_t - B_t)$ and $r_t^2 = S_t^2 + 2S_t(S_t - B_t)$.

It easily follows from Pitman's corollary, (i.e. $\rho_t = 2S_t - B_t$ and $\left(\frac{S_t}{\rho_t}\right)$ are independent, and $\left(\frac{S_t}{\rho_t}\right)$ is uniform), and from the beta–gamma algebra (Sect. 1.7), that for fixed $t > 0$, both variables $\tilde{r}_t^2$ and r_t^2 are exponentially distributed, with mean $(2t)$; moreover, r_t and $S_t - B_t$ are independent, while $\tilde{r}_t$ and S_t are independent. These remarks are due originally to Seshadri.

Since we now know Pitman's theorem, it is tempting, (but wrong!), to think that either $(r_t; t \geq 0)$ or $(\tilde{r}_t; t \geq 0)$ may be a BES(2) process. However, these processes have interesting properties:

(i) Their semimartingale decompositions are:

$$r_t^2 = -2 \int_0^t S_s dB_s + 2S_t^2$$

and

$$\tilde{r}_t^2 = -2 \int_0^t (2S_s - B_s) dB_s + (S_t^2 + t).$$

(ii) As $t \to \infty$, r_t converges to $+\infty$, since $r_t^2 > S_t^2$, while $\tilde{r}_t$ comes back indefinitely to 0, since:

$$\tilde{r}_t = (S_t - B_t)(3S_t - B_t).$$

(iii) $(r_t^2; t \geq 0)$ is a particular example of the process Φ presented in Proposition 3.6.1, i.e. $r_t^2 = R\left(4 \int_0^t du S_u^2\right)$, $t \geq 0$, where $(R(s); s \geq 0)$ is a BES(3) process.

3.8 A Combination of Skorokhod's Lemma and Time-Substitution

The following gives a further illustration of the power of Skorokhod's lemma.

Theorem 3.8.1. *Let $h > 0$. Then there exists a Brownian motion $(\beta_u; u \geq 0)$ such that, if $X_u = \beta_u + \frac{u}{2}$ and $S_u = \sup_{s \leq u} X_s$, we have the pathwise identity:*

$$\left(\log \left(1 + \frac{|B_t|}{h} \right), \frac{L_t}{h} \right) \equiv \left(S_{H_t} - X_{H_t}, S_{H_t} \right)$$

where $H_t \equiv \int_0^t \frac{ds}{(h+|B_s|)^2}$.

Proof. Develop $\left(\log \left(1 + \frac{|B_t|}{h} \right); t \geq 0 \right)$ with the help of Itô–Tanaka formula, and define $(\beta_u; u \geq 0)$ as the Dubins–Schwarz Brownian motion associated to the martingale $\left(\int_0^s \frac{\text{sgn}(B_u)dB_u}{h+|B_u|}; s \geq 0 \right)$. $\quad\square$

Exercise 3.8.2. Identify the law of $(H_{\tau_t}; t \geq 0)$, where as usual $(\tau_t; t \geq 0)$ denotes the inverse of $(L_u; u \geq 0)$.

References

1. J. Obłój, The Skorokhod embedding problem and its offspring. Probab. Surv. **1**, 321–390 (2004)
2. J. Azéma, M. Yor, Une solution simple au problème de Skorokhod. Séminaire de Probabilités, XIII (Univ. Strasbourg, Strasbourg, 1977/78). *Lecture Notes in Math.*, vol. 721. (Springer, Berlin, 1979), pp. 90–115
3. J.W. Pitman, One-dimensional Brownian motion and the three-dimensional Bessel process. Adv. Appl. Probab. **7**(3), 511–526 (1975)
4. L.C.G. Rogers, J.W. Pitman, Markov functions. Ann. Probab. **9**(4), 573–582 (1981)
5. L.C.G. Rogers, Characterizing all diffusions with the $2M - X$ property. Ann. Probab. **9**(4), 561–572 (1981)
6. Y. Saisho, H. Tanemura, Pitman type theorem for one-dimensional diffusion processes. Tokyo J. Math. **13**(2), 429–440 (1990)
7. K. Takaoka, On the martingales obtained by an extension due to Saisho, Tanemura and Yor of Pitman's theorem. Séminaire de Probabilités, XXXI. *Lecture Notes in Math.*, vol. 1655. (Springer, Berlin, 1997), pp. 256–265
8. H.P. McKean, Jr., Stochastic integrals. *Probability and Mathematical Statistics*, vol. 5. (Academic Press, New York, 1969)
9. V. Seshadri, Exponential models, Brownian motion, and independence. Can. J. Stat. **16**(3), 209–221 (1988)

Chapter 4
Paul Lévy's Arcsine Laws

We recall some representations of an arcsine distributed r.v. We decompose Brownian motion into two independent reflecting Brownian motions. We finally show that both $g_1 = \sup\{s < 1 : B_s = 0\}$ and $A_1 = \int_0^1 ds \mathbf{1}_{(B_s>0)}$ are arcsine distributed.

4.1 Two Brownian Functionals with the Arcsine Distribution

Recall that (see Sect. 1.7) a random variable A follows the arcsine law if it admits the density

$$\frac{1}{\pi\sqrt{x(1-x)}}\mathbf{1}_{[0,1)}(x).$$

Such a variable A can be represented in various ways, e.g.:

$$A \overset{\text{(law)}}{=} \frac{N^2}{N^2 + N'^2} \overset{\text{(law)}}{=} \cos^2(\theta) \overset{\text{(law)}}{=} \frac{T}{T + T'},$$

where N and N' are two independent, standard normal random variables, θ is uniform on $[0, 2\pi]$ and T and T' are two independent random variables, with the same distribution as $\frac{1}{N^2}$. Consequently,

$$E[e^{-\frac{\lambda^2}{2}T}] = e^{-\lambda}, \quad \text{for } \lambda > 0,$$

and if $T_1 = \inf\{t \geq 0 : B_t = 1\}$, then $T_1 \overset{\text{(law)}}{=} \frac{1}{N^2}$. Indeed,

J.-Y. Yen and M. Yor, *Local Times and Excursion Theory for Brownian Motion*,
Lecture Notes in Mathematics 2088, DOI 10.1007/978-3-319-01270-4_4,
© Springer International Publishing Switzerland 2013

$$(T_1 > t) = (1 > S_t) \overset{\text{(law)}}{=} (1 > \sqrt{t}\, S_1) \overset{\text{(law)}}{=} \left(\frac{1}{S_1^2} > t\right),$$

thus

$$T_1 \overset{\text{(law)}}{=} \frac{1}{S_1^2} \overset{\text{(law)}}{=} \frac{1}{B_1^2}.$$

Remark 4.1.1. To obtain the previous equalities in law, we have used the scaling property and the reflection principle for BM. More general scaling properties will be developed and exploited in Sect. 4.3.

Theorem 4.1.2 (P. Lévy). *Let* $g_1 = \sup\{t < 1 : B_t = 0\}$ *and define* $A_1^+ = \int_0^1 ds\mathbf{1}_{(B_s > 0)}$. *Then both variables follow the arc sine law.*

Proof. The law of A_1^+ will be studied later; for the moment, let us look at the law of g_1. Let $d_u = \inf\{t > u : B_t = 0\}$, then $(g_1 < u) = (1 < d_u)$, but

$$d_u = u + \inf\{v : B_{u+v} - B_u = -B_u\} = u + \hat{T}_{(-B_u)} \overset{\text{(law)}}{=} u + B_u^2 \hat{T}_1$$

$$\overset{\text{(law)}}{=} u + u B_1^2 \hat{T}_1 \overset{\text{(law)}}{=} u\left(1 + \frac{\hat{T}_1}{T_1}\right) \overset{\text{(law)}}{=} u\left(\frac{T_1 + \hat{T}_1}{T_1}\right),$$

where $(\hat{B}_v = B_{u+v} - B_u; v \geq 0)$ is a BM independent of $(B_v, v \leq u)$, and $\hat{T}_a = \inf\{v : \hat{B}_v = a\}$. Thus,

$$P(g_1 < u) = P\left(\frac{T_1}{T_1 + \hat{T}_1} < u\right).$$

Hence, $g_1 \overset{\text{(law)}}{=} \frac{T_1}{T_1 + \hat{T}_1}$, which is arc sine distributed. □

In the sequel, it will be fruitful to look for some random times T for which $\frac{1}{T}A_T^+$ follows the arc sine law and to consider random scaling of the type $(\frac{1}{\sqrt{T}} B_{uT}; u \leq 1)$.

4.2 Two Independent Reflected Brownian Motions

Consider $A_t^+ = \int_0^t \mathbf{1}_{(B_s > 0)} ds$ and $A_t^- = \int_0^t \mathbf{1}_{(B_s < 0)} ds$ the times spent up to time t in $\mathbb{R}_+$ and $\mathbb{R}_-$ respectively by $(B_s; s \leq t)$.

Theorem 4.2.1. *Let* α^+ *and* α^- *be the inverse processes of* A^+ *and* A^- *respectively. Then* $(B_{\alpha_u^+}^+; u \geq 0)$ *and* $(B_{\alpha_u^-}^-; u \geq 0)$ *are two independent reflected BM's.*

Proof. By Tanaka's formula,

$$B_t^+ = \int_0^t \mathbf{1}_{(B_s > 0)} dB_s + \frac{1}{2} L_t \quad \text{and} \quad B_t^- = -\int_0^t \mathbf{1}_{(B_s < 0)} dB_s + \frac{1}{2} L_t.$$

Knight's theorem on orthogonal martingales (see Theorem 1.3.2) implies that the martingales

$$M_t^+ = \int_0^t \mathbf{1}_{(B_s > 0)} dB_s \quad \text{and} \quad M_t^- = -\int_0^t \mathbf{1}_{(B_s < 0)} dB_s$$

satisfy

$$M_t^+ = \gamma_{A_t^+}^+ \quad \text{and} \quad M_t^- = \gamma_{A_t^-}^-,$$

where γ^+ and γ^- are two independent BM's. Moreover,

$$B_{\alpha_u^+}^+ = \gamma_u^+ + \frac{1}{2} L_{\alpha_u^+} \quad \text{and} \quad B_{\alpha_u^-}^- = \gamma_u^- + \frac{1}{2} L_{\alpha_u^-}.$$

Using Skorokhod's lemma, the processes

$$\left(\frac{1}{2} L_{\alpha_u^+}; u \geq 0\right) \quad \text{and} \quad \left(\frac{1}{2} L_{\alpha_u^-}; u \geq 0\right)$$

are the respective local times of the reflected BM's $B_{\alpha_u^+}^+$ and $B_{\alpha_u^-}^-$.

Consequently,

$$\left(\frac{1}{2} L_{\alpha_u^+}; u \geq 0\right) \overset{\text{(law)}}{=} \left(\frac{1}{2} L_{\alpha_u^-}; u \geq 0\right) \overset{\text{(law)}}{=} (L_u; u \geq 0),$$

and their inverse processes

$$\left(A_{\tau_{2t}}^+; t \geq 0\right) \quad \text{and} \quad \left(A_{\tau_{2t}}^-; t \geq 0\right),$$

where $(\tau_l, l \geq 0)$ is the inverse of $(L_t, t \geq 0)$ are independent and identically distributed as

$$(\tau_u; u \geq 0) \overset{\text{(law)}}{=} (T_u; u \geq 0).$$

$\square$

4.3   Random Brownian Scaling and Absolute Continuity Properties

It is of some interest to describe the law of $\left(\frac{1}{\sqrt{\sigma}} B_{u\sigma}, u \leq 1\right)$ for σ, a random time. In many cases, the law of this randomly scaled BM bears some absolute continuity property with that of $(B_u, u \leq 1)$.

Rather than giving a general framework right away, we start with the two following examples:

Example 4.3.1.

(a) Let $\sigma = \alpha_u^+ \equiv \inf\{t : A_t^+ > u\}$. Then

$$E\left[\mathbf{1}_{(B_1>0)} F(B_u; u \leq 1)\right] = E\left[\frac{1}{\alpha_1^+} F\left(\frac{1}{\sqrt{\alpha_1^+}} B_{u\alpha_1^+}; u \leq 1\right)\right],$$

or equivalently,

$$E\left[F\left(\frac{1}{\sqrt{\alpha_1^+}} B_{u\alpha_1^+}; u \leq 1\right)\right] = E\left[\frac{\mathbf{1}_{(B_1>0)}}{A_1^+} F(B_u; u \leq 1)\right].$$

(b) Let $\sigma = \tau_s = \inf\{t : L_t > s\}$ and define the Pseudo–Brownian bridge:

$$(B_u^{\#}; u \leq 1) \equiv \left(\frac{1}{\sqrt{\tau_s}} B_{u\tau_s}; u \leq 1\right).$$

We then have

$$E[F(b_u; u \leq 1)] = E\left[\sqrt{\frac{\pi s^2}{2\tau_s}} F(B_u^{\#}; u \leq 1)\right]$$

where $(b_u; u \leq 1)$ denotes the Brownian bridge.

For this second example, one may consult Pitman–Yor [9] and Biane–Le Gall–Yor [2].

In order to derive such absolute continuity relationships as particular cases of a general result, we consider an increasing process $(A_t; t \geq 0)$ such that $\left((B_t, A_t); t \geq 0\right)$ has the following scaling property: there exists $r \in \mathbb{R}$ such that, for each $c > 0$,

$$\left((B_{ct}, A_{ct}); t \geq 0\right) \overset{\text{(law)}}{=} \left((\sqrt{c} B_t, c^{r+1} A_t); t \geq 0\right). \tag{4.3.1}$$

Remark 4.3.2. Observe that in Example 4.3.1, property (4.3.1) is satisfied, in (a), with $A_t = A_t^+$ and $r = 0$, and in (b), $A_t = L_t$ and $r = -\frac{1}{2}$.

Coming back to our general scaling relation (4.3.1), let $\alpha_u \equiv \inf\{t : A_t > u\}$ and consider the deterministic measure $\mu^{A,F}$ on $\mathbb{R}_+$ defined by

$$I_\varphi \equiv \int_0^\infty \mu^{A,F}(dt)\varphi(t) = E\left[\int_0^\infty dA_t F\left(\frac{1}{\sqrt{t}} B_{ut}; u \leq 1\right)\varphi(t)\right],$$

or equivalently,

$$\mu^{A,F}(dt) = E\Big[dA_t\, F\Big(\frac{1}{\sqrt{t}}B_{ut}; u \le 1\Big)\Big],$$

for a functional $F : (C[0,1], \mathbb{R}) \to \mathbb{R}_+$, and a test function $\varphi : \mathbb{R}_+ \to \mathbb{R}_+$.

Theorem 4.3.3. *With the previous notation*

$$\mu^{A,F}(dt) = C_{A,F}\, t^r\, dt$$

where

$$C_{A,F} = E\Big[\frac{r+1}{\alpha_1^{r+1}}F\Big(\frac{1}{\sqrt{\alpha_1}}B_{v\alpha_1}; v \le 1\Big)\Big].$$

Proof. Consider

$$I_\varphi = E\Big[\int_0^\infty du\,\varphi(\alpha_u)F\Big(\frac{1}{\sqrt{\alpha_u}}B_{v\alpha_u}; v \le 1\Big)\Big].$$

From the scaling property (4.3.1), we deduce, for each $t > 0$, and each $u > 0$:
$A_u \overset{(\text{law})}{=} u^{r+1}A_1$ and $\alpha_u \overset{(\text{law})}{=} u^{\frac{1}{r+1}}\alpha_1$; thus the previous expression is equal to

$$I_\varphi = \int_0^\infty du\, E\Big[\varphi(u^m\alpha_1)F\Big(\frac{1}{\sqrt{\alpha_1}}B_{v\alpha_1}; v \le 1\Big)\Big] \qquad (: m = \frac{1}{r+1})$$

$$= E\Big[(r+1)\int_0^\infty dt\, t^r\, \frac{\varphi(t)}{\alpha_1^{r+1}}F\Big(\frac{1}{\sqrt{\alpha_1}}B_{v\alpha_1}; v \le 1\Big)\Big].$$

$\square$

Corollary 4.3.4. *Let $A_t = \int_0^t ds\,\theta_s$, where θ satisfies*

$$((B_{ct}, \theta_{ct}); t \ge 1) \overset{(\text{law})}{=} ((\sqrt{c}\,B_t, c^r\theta_t); t \ge 0)$$

then

$$E[\theta_1 F(B_v; v \le 1)] = E\Big[\frac{r+1}{\alpha_1^{r+1}}F\Big(\frac{1}{\sqrt{\alpha_1}}B_{v\alpha_1}; v \le 1\Big)\Big].$$

Proof. By definition

$$\mu^{A,F}(dt) = dt\, E\Big[\theta_t F\Big(\frac{1}{\sqrt{t}}B_{vt}; v \le 1\Big)\Big] = dt\, t^r\, E[\theta_1 F(B_u; u \le 1)].$$

$\square$

Remark 4.3.5. Example 4.3.1(a) is a particular case of Corollary 4.3.4.

Corollary 4.3.6. *Consider, as in Example 4.3.1(b), $A_t = L_t$. In this case,*

$$\mu^{A,F}(dt) = E\big[F(B_u; u \le 1)\big|B_1 = 0\big]\frac{dt}{\sqrt{2\pi t}}.$$

Proof. We have

$$\mu^{A,F}(dt) = E\Big[dL_t\, F(\frac{1}{\sqrt{t}}B_{ut}; u \le 1)\Big]$$

$$= E\Big[dL_t\, E[F(\frac{1}{\sqrt{t}}B_{ut}; u \le 1)\big|B_t = 0]\Big]$$

$$= E\big[F(B_u; u \le 1)\big|B_1 = 0\big]E[dL_t].$$

But

$$E[dL_t] = d_t\, E[L_t] = d_t\, E[|B_t|] = d(\sqrt{t})E[|B_1|] = \sqrt{\frac{2}{\pi}}\frac{dt}{2\sqrt{t}}.$$

$\square$

Remark 4.3.7. An important by-product of Example 4.3.1(b) is the following identity in law:

$$(B_{\tau_1-u}; u \le \tau_1) \overset{\text{(law)}}{=} (B_u; u \le \tau_1)$$

which played some part in our proof of Pitman's theorem.

4.4   The Second Arcsine Law

Theorem 4.4.1. *Let T be a random time, and consider the three-dimensional variable $Z_T \equiv \frac{1}{T}(A_T^+, A_T^-, L_T^2)$. Then Z_T has the same law in the following three cases:*

$$(i)\; T = t; \qquad (ii)\; T = \alpha_s^+; \qquad (iii)\; T = \tau_u.$$

In particular,

$$\frac{A_T^+}{T} \overset{\text{(law)}}{=} \frac{A_{\tau_u}^+}{\tau_u} \equiv \frac{A_{\tau_u}^+}{A_{\tau_u}^+ + A_{\tau_u}^-},$$

and the common law of these ratios is the arcsine distribution.

Proof. To show that (ii) and (iii) are equal in law, we have, using the definitions, the independence relations and a random time change, that

$$A^-_{\alpha^+_1} = A^-_{\tau(L_{\alpha^+_1})} \overset{(\text{law})}{=} (L_{\alpha^+_1})^2 A^-_{\tau_1}$$

$$\alpha^+_1 = A^+_{\alpha^+_1} + A^-_{\alpha^+_1} = 1 + A^-_{\alpha^+_1}$$

$$(L^2_{\alpha^+_u} < t) = (u < A^+_{\tau_{\sqrt{t}}}).$$

From this, we obtain the following identity in law

$$(A^-_{\alpha^+_1}, L^2_{\alpha^+_1}, \alpha^+_1) \overset{(\text{law})}{=} (L^2_{\alpha^+_1} A^-_{\tau_1}, L^2_{\alpha^+_1}, 1 + L^2_{\alpha^+_1} A^-_{\tau_1})$$

$$\overset{(\text{law})}{=} \left(\frac{A^-_{\tau_1}}{A^+_{\tau_1}}, \frac{1}{A^+_{\tau_1}}, 1 + \frac{A^-_{\tau_1}}{A^+_{\tau_1}} \right) = \left(\frac{A^-_{\tau_1}}{A^+_{\tau_1}}, \frac{1}{A^+_{\tau_1}}, \frac{\tau_1}{A^+_{\tau_1}} \right)$$

and from there, we also obtain:

$$\frac{1}{\alpha^+_1}(A^-_{\alpha^+_1}, L^2_{\alpha^+_1}) \overset{(\text{law})}{=} \frac{1}{\tau_1}(A^-_{\tau_1}, 1).$$

To show that (i) and (iii) are equal in law, we use the absolute continuity result in Example 4.3.1(a):

$$E[f(A^+_1, A^-_1, L^2_1)\mathbf{1}_{(B_1>0)}] = E\left[\frac{1}{\alpha^+_1} f\left(\frac{A^+_{\alpha^+_1}}{\alpha^+_1}, \frac{A^-_{\alpha^+_1}}{\alpha^+_1}, \frac{L^2_{\alpha^+_1}}{\alpha^+_1} \right) \right]$$

$$= E\left[\frac{A^+_{\tau_1}}{\tau_1} f\left(\frac{A^+_{\tau_1}}{\tau_1}, \frac{A^-_{\tau_1}}{\tau_1}, \frac{1}{\tau_1} \right) \right].$$

By symmetry, we obtain:

$$E[f(A^+_1, A^-_1, L^2_1)\mathbf{1}_{(B_1<0)}] = E\left[\frac{A^-_{\tau_1}}{\tau_1} f\left(\frac{A^+_{\tau_1}}{\tau_1}, \frac{A^-_{\tau_1}}{\tau_1}, \frac{1}{\tau_1} \right) \right].$$

Adding up the previous equalities term by term, we get:

$$E[f(A^+_1, A^-_1, L^2_1)] = E\left[f\left(\frac{A^+_{\tau_1}}{\tau_1}, \frac{A^-_{\tau_1}}{\tau_1}, \frac{1}{\tau_1} \right) \right].$$

$\square$

Comment: Clearly, in the proof of Theorem 4.4.1, computations up to the random time α_1 serve as an intermediary step to relate the computations up to fixed time (case (i)), and those up to the inverse local time (case (iii)).

In April 1995, R. Doney furthered the study up to $\alpha_1 (\equiv \alpha_1^+)$ by showing, in particular, the following result:

$$P\left(\frac{1}{g_{\alpha_1}} A_{g_{\alpha_1}}^+ \in dt\right) = \frac{dt}{2\sqrt{t}}, \quad (0 < t < 1). \tag{4.4.1}$$

Here, we propose the following reinforcement of (4.4.1):

Exercise 4.4.2. Define $\left(\tilde{b}(u) \equiv \frac{1}{\sqrt{g_{\alpha_1}}} B_{u g_{\alpha_1}}; u \leq 1\right)$, and prove that:

$$E\left[F\left(\tilde{b}(u); u \leq 1\right)\right] = E\left[\frac{1}{2\sqrt{A_b^+}} F\left(b(u); u \leq 1\right)\right]$$

where $(b(u); u \leq 1)$ is a standard Brownian bridge, and $A_b^+ = \int_0^1 du \mathbf{1}_{(b_u \geq 0)}$. Then, deduce (4.4.1) from Lemma 4.5.1 below.

4.5 The Time Spent in $\mathbb{R}_+$ by a Brownian Bridge

Lemma 4.5.1. $A_b^+ = \int_0^1 du \mathbf{1}_{(b_u \geq 0)}$ is uniformly distributed on $[0, 1]$.

Proof. The lemma is a consequence of the absolute continuity relation between b and $B^{\#}$ (see Example 4.3.1(b)). Indeed,

$$E[f(A_b^+)] = E\left[\frac{c}{\sqrt{\tau_1}} f\left(\frac{A_{\tau_1}^+}{\tau_1}\right)\right],$$

where $c = \sqrt{\frac{\pi}{2}}$.

Since $A_{\tau_1}^+$ and $A_{\tau_1}^-$ are independent and identically distributed with the same distribution as $\tau_1/4$, the expression above is equal to:

$$E\left[\frac{2c}{\left(\frac{1}{N^2} + \frac{1}{N'^2}\right)^{1/2}} f\left(\frac{\frac{1}{N^2}}{\frac{1}{N^2} + \frac{1}{N'^2}}\right)\right] = E\left[\frac{2c|N||N'|}{\sqrt{N^2 + N'^2}} f\left(\frac{N'^2}{N'^2 + N^2}\right)\right],$$

where N and N' are two centered, reduced, independent Gaussian variables. Making the change of variables:

$$N = R\cos\theta, \quad N' = R\sin\theta \quad R = \sqrt{N^2 + N'^2},$$

where θ is uniform on $[0, 2\pi]$, and R and θ are independent; we get that the previous expectation is equal to

$$2cE[R]E\big[|\cos\theta\sin\theta|f(\sin^2\theta)\big] = c'E[\sqrt{X(1-X)}f(X)],$$

with $X = \sin^2\theta$ which has the arcsine law. It follows that A_b^+ has uniform distribution on $[0,1]$. (As a check, we remark that $\frac{c'}{\pi} = \frac{2cE[R]}{\pi} = 1$).   □

4.6   The Law of A_T^+ for More Random Times T and Other Processes than BM

(a) In Sect. 4.2, and later in Sect. 4.4, we exploited the splitting of $(B_t; t \ge 0)$ into its positive (excursions) part and its negative (excursions) part with the help of Knight's theorem.

   This splitting also yields easily the joint law of, e.g.:

$$(A_{T_1}^+, A_{T_1}^-, L_{T_1}), \quad \text{where } T_1 = \inf\{t : B_t = 1\} \tag{4.6.1}$$

Theorem 4.6.1.

(1) Let $\tilde{T}_1 = \inf\{t : |B_t| = 1\}$. Then, we have:

$$(A_{T_1}^+, A_{T_1}^-, L_{T_1}) \overset{(law)}{=} (\hat{T}_1, (L_{\tilde{T}_1})^2 \hat{T}_1, 2L_{\tilde{T}_1}),$$

where $\hat{T}_1$ is independent of the pair $(\tilde{T}_1, L_{\tilde{T}_1})$, and $\hat{T}_1 \overset{(law)}{=} T_1$.
(2) The Laplace transform of the law of (4.6.1) is given by:

$$E\left[\exp\left(-\left(\frac{\lambda^2}{2}A_{T_1}^+ + \frac{\mu^2}{2}A_{T_1}^- + \alpha L_{T_1}\right)\right)\right]$$

$$= E\left[\exp\left(-\left(\frac{\lambda^2}{2}\tilde{T}_1 + (\mu + 2\alpha)L_{\tilde{T}_1}\right)\right)\right]$$

$$= \left(\cosh\lambda + \left(\frac{\mu + 2\alpha}{\lambda}\right)\sinh\lambda\right)^{-1}$$

Remark 4.6.2. This computation plays a key role in the determination of the joint law of the asymptotic small and large windings of planar BM (Pitman–Yor [8]).

(b) Instead of looking at $A_{T_1}^+$ and $A_{T_1}^-$, we now consider:

$$I_{T_1} \equiv -\inf_{s \le T_1} B_s.$$

The result:

$$P(I_{T_1} \in dx) = \frac{dx}{(1+x)^2}, \quad (x \geq 0)$$

may be obtained as a consequence of the Lemma 1.8.1. We are now interested more generally in the process

$$(I_{T_t} \equiv -\inf_{s \leq T_t} B_s; \ t \geq 0)$$

whose law may be described as follows:

Theorem 4.6.3.

(1) Define Watanabe process to be $(U(t) \equiv S_{\tau_t}; t \geq 0)$, where $S_u = \sup_{s \leq u} B_s$, and $(\tau_t; t \geq 0)$ is the inverse of the local time of B at 0. Then $(U(t); t \geq 0)$ is a Feller process.

(2) We have $(I_{T_t}; t \geq 0) \overset{(law)}{=} (U(V_t^{-1}); t \geq 0)$ where U and V are two independent Watanabe processes.

(3) The process $(Y_t = t + I_{T_t}; t \geq 0)$ is a Feller process.

(c) It is also interesting to study the occupation measure of Brownian motion with drift. More specifically, we now show how to recover from "simple" stochastic calculus Evans' formula for the Laplace transform of

$$I_{\alpha,\beta}^{(\mu)} = \int_0^\infty ds \mathbf{1}_{(\alpha < B_s + \mu s < \beta)}.$$

This consists in computing for $\alpha < \beta < a$, and $T_a = \inf\{t : X_t = a\}$, the quantity:

$$\mathbf{W}^\mu \left(\exp \left(-\lambda \int_0^{T_a} ds \mathbf{1}_{(\alpha < X_s < \beta)} \right) \right)$$

$$= \mathbf{W} \left(\exp \left(\mu a - \frac{\mu^2}{2} T_a - \lambda \int_0^{T_a} ds \mathbf{1}_{(\alpha < X_s < \beta)} \right) \right)$$

$$= \exp(\mu a) \frac{v(0)}{v(a)},$$

where $v : \mathbb{R} \to \mathbb{R}$ is a C^1-function which satisfies:

$$\frac{1}{2} v'' = \left(\lambda \mathbf{1}_{(\alpha < x < \beta)} + \frac{\mu^2}{2} \right) v.$$

Since v must also be bounded as $x \to -\infty$, it is equal to:

$$v(x) = \begin{cases} e^{\mu x}, & \text{if } x < \alpha \\ C e^{\nu x} + D e^{-\nu x}, & \text{if } \alpha < x < \beta \\ E e^{\mu x} + F e^{-\mu x}, & \text{if } x > \beta, \end{cases} \quad (: \nu = \sqrt{2\lambda + \mu^2})$$

where C, D, E, F depend on α and β, and are uniquely determined through the C^1 property of v at points α and β.

Thus, we recover, by letting $a \to \infty$, a formula obtained by S. Evans:

$$\text{for } 0 < \alpha < \beta, \quad E\left[\exp\left(-\lambda I_{\alpha,\beta}^{(\mu)}\right)\right] = \frac{1}{E_{\alpha,\beta}},$$

where

$$E_{\alpha,\beta} = \frac{(\mu + \nu)^2}{4\mu\nu} \exp\left((\nu - \mu)(\alpha - \beta)\right) - \frac{(\mu - \nu)^2}{4\mu\nu} \exp\left(-(\nu + \mu)(\alpha - \beta)\right),$$

with $\nu = \sqrt{\mu^2 + 2\lambda}$. Note that:

$$\frac{1}{E_{\alpha,\beta}\left(\frac{\lambda}{\beta - \alpha}\right)} \xrightarrow[\beta \downarrow \alpha]{} E\left[\exp\left(-\lambda L_\infty^\alpha\left(B^{(\mu)}\right)\right)\right].$$

References

1. M. Barlow, J. Pitman, M. Yor, Une extension multidimensionnelle de la loi de l'arc sinus. Séminaire de Probabilités, XXIII. *Lecture Notes in Math.*, vol. 1372. (Springer, Berlin, 1989), pp. 294–314
2. Ph. Biane, J.-F. Le Gall, M. Yor, Un processus qui ressemble au pont brownien. Séminaire de Probabilités, XXI. *Lecture Notes in Math.*, vol. 1247. (Springer, Berlin, 1987), pp. 270–275
3. S.N. Evans, Multiplicities of a random sausage. Ann. Inst. H. Poincaré Probab. Stat. **30**(3), 501–518 (1994)
4. T. Jeulin, M. Yor, Sur les distributions de certaines fonctionnelles du mouvement brownien. Seminar on Probability, XV (Univ. Strasbourg, Strasbourg, 1979/1980) (French). *Lecture Notes in Math.*, vol. 850. (Springer, Berlin, 1981), pp. 210–226
5. Y. Kasahara, Y. Yano, On a generalized arc-sine law for one-dimensional diffusion processes. Osaka J. Math. **42**(1), 1–10 (2005)
6. P. Lévy, Sur certains processus stochastiques homogènes. Compositio Math. **7**, 283–339 (1939)
7. H.P. McKean, Brownian local times. Adv. Math. **16**, 91–111 (1975)
8. J. Pitman, M. Yor, Asymptotic laws of planar Brownian motion. Ann. Probab. **14**(3), 733–779 (1986)
9. J. Pitman, M. Yor, Arcsine laws and interval partitions derived from a stable subordinator. Proc. Lond. Math. Soc. (3), **65**(2), 326–356 (1992)
10. M. Yor, Random Brownian scaling and some absolute continuity relationships. Seminar on Stochastic Analysis, Random Fields and Applications (Ascona, 1993). *Progr. Probab.*, vol. 36. (Birkhäuser, Boston, 1995), pp. 243–252

11. S. Watanabe, Generalized arc-sine laws for one-dimensional diffusion processes and random walks. Stochastic analysis (Ithaca, NY, 1993). Proc. Sympos. Pure Math. **57**, 157–172 (1995)
12. S. Watanabe, K. Yano, Y. Yano, A density formula for the law of time spent on the positive side of one-dimensional diffusion processes. J. Math. Kyoto Univ. **45**(4), 781–806 (2005)
13. D. Williams, Markov properties of Brownian local time. Bull. Am. Math. Soc. **75**, 1035–1036 (1969)

Part II
Excursion Theory for Brownian Paths

Chapter 5
Brownian Excursion Theory: A First Approach

Itô's excursion process associated to Brownian motion is presented, along with master additive and multiplicative formulae for excursion theory.

5.1 Some Motivations

Brownian excursion theory will allow us to get deeper into several studies of functionals of BM, for which a given level, 0 say, plays some "distinguished" role.

Example 5.1.1.

(a) Consider the Brownian additive functional associated to a function $f \in L^1_{loc}(\mathbb{R})$:

$$A_t^{(f)} = \int_0^t du\, f(B_u),\quad t \geq 0.$$

Then, $(A_{\tau_l}^{(f)}; l \geq 0)$ is a Lévy process, i.e., a process with homogeneous independent increments. More generally, the processes

$$\left(A_{\tau_l}^{(\nu)} \equiv \int_{-\infty}^{\infty} L_{\tau_l}^x \nu(dx); \ l \geq 0 \right)$$

belong to an interesting subclass of Lévy processes, where $\nu(dx)$ is a Radon measure; indeed, it can be deduced from Krein's theory of strings, that the Lévy measure $m(da)$ of every such process is of the form: $m(da) = \mu(a)da$, where μ is the Laplace transform of a positive measure.
(b) Principal values of Brownian local times.
As an extension case of part (a), consider, for $\alpha < 3/2$:

J.-Y. Yen and M. Yor, *Local Times and Excursion Theory for Brownian Motion*, Lecture Notes in Mathematics 2088, DOI 10.1007/978-3-319-01270-4_5, © Springer International Publishing Switzerland 2013

$$H^{(\alpha)}_{\tau_l} = \int_0^{\tau_l} \frac{ds}{B^\alpha_s}, \quad (\text{here, } x^\alpha \equiv |x|^\alpha \text{sgn}(x)).$$

This class of Lévy processes is particularly interesting in connection with stable processes.

Remark 5.1.2. Convergence questions involving principal values of Brownian local times have been considered in Sect. 2.6.

Proposition 5.1.3. $(H^{(\alpha)}_{\tau_l}; l \geq 0)$ *is a symmetric stable process of index* $\gamma = \frac{1}{2-\alpha}$. *Consequently,*

$$E[\exp(i\lambda H^{(\alpha)}_{\tau_l})] = \exp(-c_\alpha l |\lambda|^\gamma).$$

Proof. A change of variables and the scaling property for BM justify the following calculation

$$\int_0^{\tau_{cl}} \frac{ds}{B^\alpha_s} = c^2 \int_0^{\frac{1}{c^2}\tau_{cl}} \frac{dv}{B^\alpha_{c^2 v}} \stackrel{(\text{law})}{=} c^{2-\alpha} \int_0^{\tau_l} \frac{dv}{B^\alpha_v}$$

i.e.

$$H^{(\alpha)}_{\tau_{cl}} = c^{2-\alpha} H^{(\alpha)}_{\tau_l}$$

and since $(H^{(\alpha)}_{\tau_l}; l \geq 0)$ is a symmetric Lévy process, there exists a function ψ : $\mathbb{R}_+ \to \mathbb{R}$ such that

$$E[\exp(i\lambda H^{(\alpha)}_{\tau_l})] = \exp(-l\psi(|\lambda|)).$$

Using the scaling property for $H^{(\alpha)}_{\tau_l}$, we get:

$$E[\exp(i\lambda H^{(\alpha)}_{\tau_l})] = E[\exp(i\lambda l^{2-\alpha} H^{(\alpha)}_{\tau_1})] = \exp(-\psi(|\lambda| l^{2-\alpha})).$$

Hence, $l\psi(|\lambda|) = \psi(|\lambda| l^{2-\alpha})$, so $l\psi(1) = \psi(l^{2-\alpha})$, and we get

$$\psi(l) = c_\alpha l^{1/(2-\alpha)}$$

which finishes the proof. $\square$

Remark 5.1.4. The function $\gamma(\alpha) = \frac{1}{2-\alpha}$ transforms the interval $]-\infty, 3/2[$ of convergence of the functionals $H^{(\alpha)}_{\tau_l}$, onto the interval $]0, 2[$, i.e., every symmetric stable process of index $\gamma \in]0, 2[$ may be realized as a process $H^{(\alpha)}_{\tau_l}$ with $\gamma = \frac{1}{2-\alpha}$.

One often considers realizations of symmetric stable processes in terms of two independent BM's. A simple example is the representation of the Cauchy process $(C_l; l \geq 0)$ as

$$C_l \overset{\text{(law)}}{=} \hat{B}_{\tau_l} \overset{\text{(law)}}{=} \hat{B}_{T_l}$$

where $T_l = \inf\{s : B_s = l\}$, B and $\hat{B}$ are two independent BM's. Thus, it is remarkable that $(H_u^{(1)}; u \geq 0)$ restricted to the set $(\tau_l; l \geq 0)$ of zeros of B behaves as a BM independent of $(\tau_l; l \geq 0)$.

Remark 5.1.5. Once the basics of excursion theory have been presented in Sect. 5.2, it shall become clear that the study of the Lévy process $(A_{\tau_l}^{(v)}, l \geq 0)$ say shall be facilitated by excursion theory.

5.2 Itô's Theorem on Excursions

Define $\mathcal{Z}_\omega \equiv \{t : B_t(\omega) = 0\}$ and let $(\tau_l; l \geq 0)$ be the inverse local time. The complement of $\mathcal{Z}_\omega$ is shown to be:

$$\mathcal{Z}_\omega^c = \bigcup_{s>0}]\tau_{s-}, \tau_s[$$

i.e. where $]\tau_{s-}, \tau_s[$ are the maximal intervals of constancy of the local time L.
    We define the excursion process as

$$\mathbf{e}_l(\omega)(t) = B_{(\tau_{l-}+t)} \mathbf{1}_{(t \leq \tau_l - \tau_{l-})}, \quad (l \geq 0).$$

This is a path-valued process $\mathbf{e} : \mathbb{R}_+ \to \Omega_*$, where

$$\Omega_* = \{\varepsilon : \mathbb{R}_+ \to \mathbb{R} : \exists V(\varepsilon) < \infty, \text{ with } \varepsilon(V(\epsilon) + t) \equiv 0, \forall t \geq 0,$$

$$\varepsilon(u) \neq 0, \forall 0 < u < V(\varepsilon), \varepsilon(0) = 0, \varepsilon \text{ is continuous}\}$$

We shall denote by 0^* the excursion ε such as $V(\varepsilon) = 0$, and we define $\mathcal{F}_t^* = \sigma\{\varepsilon(u), u \leq t\}, t \geq 0$ the canonical filtration on Ω_*.

Warning. The term "excursion" is often used with either one of the three different meanings; it may indicate:

(i) *the excursion process* $(\mathbf{e}_l; l \geq 0)$, or
(ii) *the generic excursion* $\varepsilon \in \Omega_*$, or
(iii) *a particular excursion*, e.g. $(B_{g_T+u}; u \leq d_T - g_T)$, i.e. the so-called excursion straddling time T.

As usual with such abuses of language, the context often helps to clarify which kind of "excursion" is being considered.

Theorem 5.2.1 (Itô). $(e_s, s \geq 0)$ *is a Poisson Point Process. More precisely it is an* $(\mathcal{F}_{\tau_s}, s \geq 0)$-*PPP, i.e. if* Γ *is a measurable set in* Ω_* *and*

$$N_l^{\Gamma} \equiv \sum_{s \leq l} 1_{(e_s \in \Gamma)}$$

then $(N_l^{\Gamma}; l \geq 0)$ *is either identically infinite or it is a Poisson process with respect to* $(\mathcal{F}_{\tau_l})$. *Moreover, if* $\Gamma_1, \Gamma_2, \ldots, \Gamma_k$ *are disjoint, then* $N^{\Gamma_1}, N^{\Gamma_2}, \ldots, N^{\Gamma_k}$ *are independent Poisson processes.*

Proof. Straightforward, as a consequence of the Strong Markov property at τ_l.   □

If $N_l^{\Gamma} < \infty$ we call $\mathbf{n}(\Gamma)$ the intensity of $(N_l^{\Gamma}; l \geq 0)$, i.e. the positive real μ such that $(N_l^{\Gamma} - l\mu; l \geq 0)$ is an $(\mathcal{F}_{\tau_l})$-martingale.

The application $\Gamma \to \mathbf{n}(\Gamma)$ is σ-additive on Ω_*; hence, $\mathbf{n}$ is a measure which we shall describe below.

5.3   Two Master Formulae (A) and (M)

(1) *The additive formula.* Let G be a $\mathcal{P} \otimes \mathcal{F}_*$-measurable functional ($\mathcal{P}$ is the $(\mathcal{F}_t)$ predictable σ-algebra on $\mathbb{R}_+ \times \Omega$). Then

$$E\left[\sum_{s>0} G(\tau_{s-}, \omega; e_s)\right] \overset{(a)}{=} E\left[\int_0^{\infty} ds \int_{\Omega_*} \mathbf{n}(d\varepsilon) G(\tau_{s-}, \omega; \varepsilon)\right] \qquad \text{(A)}$$

$$\overset{(b)}{=} E\left[\int_0^{\infty} dL_u \int_{\Omega_*} \mathbf{n}(d\varepsilon) G(u, \omega; \varepsilon)\right]$$

where we assume that $G(u, \omega; 0^*) = 0$.

Remark 5.3.1. On the RHS of (A), one can replace τ_{s-} by τ_s since one integrates with respect to Lebesgue measure ds.

Example 5.3.2.

$$\int_0^{\infty} dt \exp\left(-\lambda t - \int_0^t du\, f(B_u)\right) = \sum_{s>0} \int_{\tau_{s-}}^{\tau_s} dt \exp\left(-\lambda t - \int_0^t du\, f(B_u)\right).$$

Proof of (A). By the Monotone Class Theorem, it suffices to consider functionals G of the form $G(t, \omega; \varepsilon) = g(t, \omega) 1_{\Gamma}(\varepsilon)$. Thus

$$E\left[\sum_{s>0} g(\tau_{s-},\omega)\mathbf{1}_\Gamma(\mathbf{e}_s)\right] = E\left[\int_0^\infty g(\tau_{s-},\omega)dN_s^\Gamma\right]$$

$$= E\left[\mathbf{n}(\Gamma)\int_0^\infty ds\; g(\tau_{s-},\omega)\right]$$

from the definition of **n**.

(2)   *The multiplicative formula.* Let $f : \mathbb{R}_+ \times \Omega_* \to \mathbb{R}_+$ be measurable with $f(s,0^*)=0$. Then

$$E\left[\exp\left(-\sum_{s>0} f(s,\mathbf{e}_s)\right)\right] = \exp\left[-\int_0^\infty ds \int_{\Omega_*}\mathbf{n}(d\varepsilon)\big(1-\exp(-f(s,\varepsilon))\big)\right]$$

$$\text{(M)}$$

Proof. Let $(\Gamma_i)_{1\le i\le k}$ be disjoint sets and consider the function $f(s,\varepsilon) = \sum_{i=1}^k f_i(s)\mathbf{1}_{\Gamma_i}(\varepsilon)$, then

$$\sum_{s>0} f(s,\mathbf{e}_s) = \sum_{i=1}^k \int_0^\infty f_i(s)dN_s^{\Gamma_i}$$

and since the N^{Γ_i}'s are independent Poisson processes, (M) is the usual exponential formula for Poisson processes.   □

5.4   Relationship Between Certain Lévy Measures and Itô Measure n

Consider functionals as in Example 5.1.1(a):

$$A_{\tau_l}^{(f)} = \int_0^{\tau_l} ds\; f(B_s).$$

For $f \ge 0$, the multiplicative formula (M) in Sect. 5.3 implies

$$E[\exp(-\lambda A_{\tau_l}^{(f)})] = \exp\left[-l\int_{\Omega_*}\mathbf{n}(d\varepsilon)\big(1-\exp(-\lambda\int_0^V ds\; f(\varepsilon_s))\big)\right].$$

On the other hand, from the definition of the Lévy measure m_f associated to the Lévy process $(A_{\tau_l}^{(f)}; l \ge 0)$, we have

$$E[\exp(-\lambda A_{\tau_l}^{(f)})] = \exp\left(-l\int m_f(dx)\big(1-\exp(-\lambda x)\big)\right).$$

Hence, $m_f(dx)$ is the image of $\mathbf{n}$ by the application

$$\varepsilon \to \int_0^{V(\varepsilon)} du\, f(\varepsilon_u).$$

(a) In particular, for:

$$A_u^{(f)} \equiv A_u^+ \equiv \int_0^u ds \mathbf{1}_{(B_s > 0)}$$

Remark 5.3.1 gives an elementary verification of the independence of the processes $(A_{\tau_l}^+; l \geq 0)$ and $(A_{\tau_l}^-; l \geq 0)$. Indeed, taking $f(x) = \lambda \mathbf{1}_{(x>0)} + \mu \mathbf{1}_{(x<0)}$ we get

$$\int_{\Omega_*} \mathbf{n}(d\varepsilon)\Big(1 - \exp\Big(-\int_0^V ds\, f(\varepsilon_s)\Big)\Big) = \int_{\Omega_*^+} \mathbf{n}(d\varepsilon)\big(1 - \exp(-\lambda V)\big)$$

$$+ \int_{\Omega_*^-} \mathbf{n}(d\varepsilon)\big(1 - \exp(-\mu V)\big)$$

whence we obtain a relation for the joint Laplace transform of $A_{\tau_l}^+$ and $A_{\tau_l}^-$ which shows the independence.

(b) Let now $f(x) = \frac{1}{x^\alpha}$ with $\alpha < 3/2$, then

$$E\Big[\exp\Big(i\lambda \int_0^{\tau_l} ds\, f(B_s)\Big)\Big] = \exp\Big(-l \int \mathbf{n}(d\varepsilon)\Big(1 - \exp\Big(i\lambda \int_0^V ds\, f(\varepsilon_s)\Big)\Big)\Big).$$

Comparing with Fourier transform obtained in Example 5.1.1(b), we get

$$c_\alpha |\lambda|^{\gamma(\alpha)} = \int \mathbf{n}(d\varepsilon)\Big(1 - \exp\Big(i\lambda \int_0^V ds\, f(\varepsilon_s)\Big)\Big),$$

where $\gamma(\alpha) = 1/(2 - \alpha)$ as in Remark 5.1.4.

We can obtain the same result using, for example, the first Itô–Williams description of $\mathbf{n}$ in Chap. 6.

5.5   Two Applications of (A) and (M)

(i) Take $f \equiv 1$ in Sect. 5.4. We know (from Lévy's equivalence theorem, for instance) that:

$$(\tau_l; l \geq 0) \stackrel{(\text{law})}{=} (T_l, l \geq 0),$$

where $T_l = \inf\{t : B_t > l\}$. Consequently,

$$E[\exp(-\lambda \tau_l)] = \exp(-l\sqrt{2\lambda}),$$

and it is easily shown, with the notation of Sect. 5.4 that:

$$m_1(dv) \equiv \mathbf{n}(V \in dv) = \frac{dv}{\sqrt{2\pi v^3}}.$$

(ii) A slight extension of (M) yields:

$$\mathbf{W}(S_{\tau_l} \le m) = \exp(-l\,\mathbf{n}(M \ge m)).$$

On the other hand, we have:

$$(S_{\tau_l} \le m) = (l < L_{T_m})$$
$$\overset{(\text{law})}{=} \left(\frac{l}{m} < L_{T_1}\right).$$

Recalling that $\frac{1}{2}L_{T_1}$ is a standard exponential, we recover:

$$P(S_{\tau_l} \le m) = \exp\left(-\frac{l}{2m}\right),$$

so that, in terms of $\mathbf{n}$, we get, for $M(\varepsilon) = \sup_{s \le V}\varepsilon(s)$:

$$\mathbf{n}(M \ge m) = \frac{1}{2m},$$

i.e.

$$\mathbf{n}_+(M \in dm) = \frac{dm}{2m^2}.$$

Pushing the study of $(S_{\tau_l})_{l \ge 0}$ at the process level, it is easily shown that $(S_{\tau_l}; l \ge 0)$ is a strong Markov process with semigroup:

$$P_l f(\sigma) = f(\sigma)\exp\left(-\frac{l}{2\sigma}\right) + \int_\sigma^\infty \exp\left(-\frac{l}{2m}\right)\frac{l}{2m^2} f(m)\,dm.$$

Comment: For a more extended survey of Itô's excursion theory, see Pitman–Yor [6].

References

1. K. Itô, Poisson point processes attached to Markov processes, in *Proceedings of the Sixth Berkeley Symposium on Mathematical Statistics and Probability (Univ. California, Berkeley, Calif., 1970/1971)*, Vol. III: Probability theory. Univ. (California Press, Berkeley, 1972), pp. 225–239
2. P.A. Meyer, Processus de Poisson ponctuels, d'après K. Itô. Séminaire de Probabilités, V (Univ. Strasbourg, année universitaire 1969–1970). *Lecture Notes in Math.*, vol. 191. (Springer, Berlin, 1971), pp. 177–190
3. J. Bertoin, Applications de la théorie spectrale des cordes vibrantes aux fonctionnelles additives principales d'un brownien réfléchi. Ann. Inst. H. Poincaré Probab. Stat. **25**(3), 307–323 (1989)
4. S. Kotani, S. Watanabe, Kreĭn's spectral theory of strings and generalized diffusion processes. Functional analysis in Markov processes (Katata/Kyoto, 1981). *Lecture Notes in Math.*, vol. 923. (Springer, Berlin, 1982), pp. 235–259
5. F.B. Knight, Characterization of the Levy measures of inverse local times of gap diffusion. Seminar on Stochastic Processes, 1981 (Evanston, Ill., 1981). *Progr. Prob. Statist.*, vol. 1. (Birkhäuser, Boston, 1981), pp. 53–78
6. J. Pitman, M. Yor, Itô's excursion theory and its applications. Jpn. J. Math. **2**(1), 83–96 (2007)

Chapter 6
Two Descriptions of n: Itô's and Williams'

We discuss two disintegrations of the Itô measure **n**, respectively in terms of the lifetime V and the maximum M.

This chapter, and more generally, all discussions pertaining to excursion theory, is a condensed version of Chap. XII in Revuz–Yor [2], where the reader shall find more details.

6.1 Statements

The next two descriptions of **n** are extremely useful. Their importance justifies that we put them in boxes:

---

(1) (Itô's representation of **n**)
Conditioning with respect to the lifetime V.

(a) The "law" of V under **n** is:

$$\mathbf{n}(V \in dv) = \frac{dv}{\sqrt{2\pi v^3}};$$

(b) Conditionally on $V = v$, $(|\varepsilon_t|; t \le v)$ is a BES(3) bridge with length v.

Remark 6.1.1. From the direct sum decomposition $\Omega_* = \Omega_*^+ \cup \Omega_*^-$, we can take $\mathbf{n} = \mathbf{n}_+ + \mathbf{n}_-$, and write the adequate versions of (1) and (2) (see the next statement) for $\mathbf{n}_+$ and $\mathbf{n}_-$.

(2) (Williams' representation of **n**)

Conditioning with respect to $M(\varepsilon) \equiv \max_{s \leq V} |\varepsilon_s|$.

(a) The "law" of M under **n** is:

$$\mathbf{n}(M \in dm) = \frac{dm}{m^2};$$

(b) Conditionally on $M = m$, $(|\varepsilon_t|; t \leq T_m)$ and $(|\varepsilon_{V-t}|; t \leq V - T_m)$ are two independent BES(3) processes considered up to their first hitting times of level m.

We already know parts (a) of each of these descriptions (see Sect. 5.5); thus, we shall concentrate on proving parts (b).

Before doing this, let us show how, thanks to (2) above, we have access, on one hand, to the joint law of (S_{τ_l}, τ_l), and on the other hand, to the joint "law" under **n** of (S_V, V).

Again, a slight extension of (M) allows to write:

$$E_W\left[\mathbf{1}_{(S_{\tau_l} \leq m)} \exp\left(-\frac{\xi^2}{2}\tau_l\right)\right] = \exp\left(-l\int \mathbf{n}(d\varepsilon)\left(1 - \mathbf{1}_{(S_V \leq m)} \exp\left(-\frac{\xi^2}{2}V\right)\right)\right).$$

Conditioning with respect to S_V, and using (2) above, we get:

$$\mathbf{n}_+\left(1 - \exp\left(-\frac{\xi^2}{2}V\right)\Big|S_V = \theta\right) = 1 - \left(\frac{\theta\xi}{\sinh(\theta\xi)}\right)^2.$$

A simple computation shows that:

$$\frac{1}{2m} + \frac{\xi}{2}\int_0^{m\xi} \frac{d\theta}{\theta^2}\left(1 - \left(\frac{\theta}{\sinh\theta}\right)^2\right) = \frac{\xi\coth(m\xi)}{2} \qquad (6.1.1)$$

so that we may characterize the joint law of (S_{τ_l}, τ_l) by:

$$E_W\left[\mathbf{1}_{(S_{\tau_l} \leq m)} \exp\left(-\frac{\xi^2}{2}\tau_l^+\right)\right] = \exp\left(-l\frac{\xi\coth(m\xi)}{2}\right).$$

Remark 6.1.2. If we let $m \to \infty$ in (6.1.1), we recover the remarkable fact that

$$\int_0^\infty \frac{d\theta}{\theta^2}\left(1 - \left(\frac{\theta}{\sinh\theta}\right)^2\right) = 1.$$

Thus, $\frac{1}{\theta^2}\left(1 - \left(\frac{\theta}{\sinh\theta}\right)^2\right)$ is a probability density.

Question: of which remarkable r.v.?

6.2   An Agreement Formula

Before proving parts (b) described in Sect. 6.1, we note that for the two descriptions to be true simultaneously, there must exist some relationship between the standard BES(3) bridge (Itô) and the process obtained by putting two BES(3) back to back (Williams). We call this relationship an agreement formula.

Here is a first attempt; in particular, we have:

$$\int \mathbf{n}(d\varepsilon) f(V) F\left(\frac{\varepsilon(uV)}{\sqrt{V}}, u \le 1\right)$$

$$\equiv \int \frac{dv}{\sqrt{2\pi v^3}} f(v) \Pi\left(F(r(u), u \le 1)\right), \quad \text{on one hand,}$$

where Π denotes the law of the standard BES(3) bridge $(r(u), u \le 1)$,

$$= \int \mathbf{n}(d\varepsilon) f(T_M + T_M') E\left[F\left(\frac{\hat{r}((T_M + T_M')u)}{\sqrt{T_M + T_M'}}, u \le 1\right)\right], \quad \text{on the other hand,}$$

where $\hat{r}$ is obtained by putting the two BES(3) processes back to back, as described at the end of Williams' representation of $\mathbf{n}$.

We now look at a particular case, that is, in this section, we only look at an example, but this is the most important (probably). It will then be conveniently taken up and extended in Corollary 11.6.3 for all $d \ge 2$.

$$\int \mathbf{n}(d\varepsilon) f(V) \Pi\left(F\left(\sup_{u \le 1}(r(u))^2\right)\right) = \int \mathbf{n}(d\varepsilon) E\left[f(T_M + T_M') F\left(\frac{1}{T_M + T_M'} M^2\right)\right].$$

By scaling and the knowledge of the laws of V and T_M, we get:

$$\left(\int_0^\infty \frac{dv}{\sqrt{2\pi v^3}} f(v)\right) \Pi\left(F\left(\sup_{u \le 1}(r(u))^2\right)\right)$$

$$= \int \frac{dm}{m^2} E\left[f(m^2 \Sigma) F\left(\frac{m^2}{m^2 \Sigma}\right)\right] \tag{6.2.1}$$

where $\Sigma = T_1 + T_1'$.

By change of variables: $v = m^2 \Sigma$ for the second integral in (6.2.1), we write: $m = \sqrt{\frac{v}{\Sigma}}, dm = \frac{dv}{2\sqrt{v\Sigma}}$. Then the integral becomes:

$$E\left[\int \frac{dv}{2\sqrt{v\Sigma}} \frac{1}{\left(\frac{v}{\Sigma}\right)} f(v) F\left(\frac{1}{\Sigma}\right)\right] = \frac{1}{2} \int \frac{dv f(v)}{v^{3/2}} E\left[\sqrt{\Sigma} F\left(\frac{1}{\Sigma}\right)\right].$$

Finally, returning to (6.2.1), we obtain:

$$\Pi\left(F\left(\sup_{u\leq1}(r(u))^2\right)\right) = \sqrt{\frac{\pi}{2}}\, E\left[\sqrt{\Sigma}\, F\left(\frac{1}{\Sigma}\right)\right].$$

For a more complete agreement formula, see Corollary 11.6.3, and even better Theorem 11.6.1

6.3   n is a Markovian Measure

Precisely:

Theorem 6.3.1. *Under* **n**, *the process* $(\varepsilon_t; t > 0)$ *is a homogeneous Markov process with:*

(i) its semigroup transition being Q_t, the semigroup of BM killed at 0;
(ii) entrance laws $(\lambda_t; t > 0)$ being:

$$\lambda_t(dy) = l_t(|y|)dy, \quad with$$

$$l_t(a) = \frac{1}{\sqrt{2\pi t^3}}\, a\, \exp\left(-\frac{a^2}{2t}\right), \quad (a > 0)$$

Recall that $(l_t(a); t > 0)$ is the density of $T_a = \inf\{t : B_t = a\}$. Thus, from (i) and (ii), there is the relationship:

$$\lambda_t Q_s = \lambda_{t+s}.$$

We shall now proceed to prove both *(i)* and *(ii)*, but some preliminary work needs to be done.

6.4   Proof of Itô's Disintegration (b) in Sect. 6.1

For now, we assume Theorem 6.3.1, which shall be fully proven later.

(b) *Itô's disintegration theorem.* We wish to show

$$\mathbf{n}_+(\Gamma) = \frac{1}{2} \int_0^\infty \frac{dr}{\sqrt{2\pi r^3}}\, \Pi^r(\Gamma) \tag{6.4.1}$$

where Π^r denotes the law of the BES(3) bridge of duration r.
We shall verify identity (6.4.1) by monotone class, starting with

$$\Gamma = \bigcap_{i=1}^{n} \Big(\varepsilon(t_i) \in A_i \Big), \quad t_1 < \cdots < t_n,$$

where $A_i \in]0, \infty[$, for all i. Note that on that set Γ, we have: $t_n < V$.
To obtain formula (6.4.1), it stems from Markov property, on the left hand side
that we have :

$$\mathbf{n}_+(\Gamma) = \int_{A_1} l_{t_1}(x_1)dx_1 \int_{A_2} q_{t_2-t_1}(x_1, x_2)dx_2 \cdots \cdots \tag{6.4.2}$$

$$\times \int_{A_n} q_{t_n-t_{n-1}}(x_{n-1}, x_n)dx_n.$$

On the other hand, we need to calculate $\Pi^r(\Gamma)$. We assume $t_n < r$, and recall
that

$$\Pi^r(V < t_n) = 0. \tag{6.4.3}$$

It is not difficult to show directly the formula for the Bessel bridge:

$$\Pi^r(\Gamma) = \int_{A_1} 2\sqrt{2\pi r^3}\, l_{t_1}(x_1)dx_1 \int_{A_2} q_{t_2-t_1}(x_1, x_2)dx_2 \cdots$$

$$\times \int_{A_n} q_{t_n-t_{n-1}}(x_{n-1}, x_n)l_{r-t_n}(x_n)dx_n. \tag{6.4.4}$$

To obtain Itô's Theorem, it remains for us to integrate $\int_{t_n}^{\infty} \frac{dr}{2\sqrt{2\pi r^3}}\Pi^r(\Gamma)$; from
(6.4.4), and simplifying the term $2\sqrt{2\pi r^3}$, we obtain:

$$\int_{t_n}^{\infty} dr l_{r-t_n}(x_n) = 1,$$

as has already been noted. Thus, we recover formula (6.4.2) hence (6.4.1).
It remains to demonstrate:
(c) the Markovian character of $\mathbf{n}$;
(d) formula (6.4.4) for $\Pi^r(\Gamma)$, which will clearly follow from the semigroup
expression of the BES(3) process.

6.5   Proof of the Formula (6.4.4) for $\Pi^r(\Gamma)$

Recall that $(P_t^{(3)})$ is a h-transform of Q_t

$$P_t^{(3)} f(y) = \frac{1}{y} Q_t(fx)(y).$$

This shall enable us to give a formula for the marginals of the bridge $\Pi_{0,z}^r$ ($z \neq 0$), and we shall later let $z \to 0$.

We have, for $t_1 < t_2 < \cdots < t_n < r$, and generic functions f and φ:

$$E_0^{(3)}[f(X_{t_1}, \cdots, X_{t_n})\varphi(X_r)] = \int p_r^{(3)}(0, z)dz\varphi(z)\Pi_{0,z}^r(f(X_{t_1}, \cdots, X_{t_n}))$$

Applying the Markov property at time t_n, we get:

$$E_0^{(3)}\left[f(X_{t_1}, \cdots, X_{t_n}) P_{r-t_n}^{(3)}\varphi(X_{t_n})\right] = E_0^{(3)}\left[f(X_{t_1}, \cdots, X_{t_n})\right.$$
$$\left. \times \frac{1}{X_{t_n}} \int q_{r-t_n}(X_{t_n}, z)z\varphi(z)dz\right].$$

Thus

$$\Pi_{0,z}^r(f(X_{t_1}, \cdots, X_{t_n})) = E_0^{(3)}\left[f(X_{t_1}, \cdots, X_{t_n})\frac{1}{X_{t_n}}\frac{q_{r-t_n}(X_{t_n}, z)z}{p_r^{(3)}(0, z)}\right].$$

It remains to let $z \to 0$ to obtain:

$$q_{r-t_n}(x_n, z)\frac{z}{p_r^{(3)}(0, z)} \xrightarrow[(z\to 0)]{} 2\, l_{r-t_n}(x_n)\sqrt{2\pi r^3}, \qquad (6.5.1)$$

The details are left as Exercise 6.5.1

Exercise 6.5.1. Prove the convergence (6.5.1).

Detailed hint:

$$p_r^{(3)}(0, z) = \frac{1}{\sqrt{2\pi r^3}}z^2 e^{-z^2/2r}$$

$$\sim \frac{z^2}{\sqrt{2\pi r^3}} \quad \text{(as } z \to 0\text{)}.$$

On the other hand,

$$q_{r-t_n}(x_n, z) = \frac{\exp\left(-\frac{(x_n-z)^2}{2(r-t_n)}\right) - \exp\left(-\frac{(x_n+z)^2}{2(r-t_n)}\right)}{\sqrt{2\pi(r-t_n)}}$$

$$\underset{z\to 0}{\sim} \frac{1}{\sqrt{2\pi(r-t_n)}} \exp\left(-\frac{x_n^2}{2(r-t_n)}\right) \frac{2x_n z}{(r-t_n)}. \qquad (6.5.2)$$

Thus, going back to (6.5.1), the LHS converges to

$$\exp\left(-\frac{x_n^2}{2(r-t_n)}\right) \sqrt{2\pi r^3} \frac{2x_n}{(r-t_n)} \frac{1}{\sqrt{2\pi(r-t_n)}} \equiv 2l_{r-t_n}(x_n)\sqrt{2\pi r^3},$$

and (6.5.1) is completely proven.

Thus, we have obtained the formula:

$$\Pi_{0,0}^r\big(f(X_{t_1}, \cdots X_{t_n})\big) = E_0^3\Big[f(X_{t_1}, \cdots X_{t_n})\frac{2}{X_{t_n}}l_{r-t_n}(X_{t_n})\sqrt{2\pi r^3}\Big] \qquad (6.5.3)$$

to be compared with formula (6.4.4):

$$\Pi_{0,0}^r\big(f(X_{t_1}, \cdots X_{t_n})\big) = \int\int \cdots \int f(x_1, \cdots, x_n)dx_1 \cdots$$

$$\times dx_n 2\sqrt{2\pi r^3}l_{t_1}(x_1)q_{t_2-t_1}(x_1, x_2) \cdots$$

$$\times q_{t_n-t_{n-1}}(x_{n-1}, x_n)l_{r-t_n}(x_n).$$

We write the semigroup $P_t^{(3)}(x, dy) = p_t^{3\uparrow}(x, y)y^2 dy$ since $y^2 dy$ is the speed measure of BES(3). Hence, the comparison of (6.5.3) and (6.4.4) now boils down to:

$$p_{t_1}^{3\uparrow}(0, x_1)x_1^2 \, p_{t_2-t_1}^{3\uparrow}(x_1, x_2)x_2^2 \cdots p_{t_n-t_{n-1}}^{3\uparrow}(x_{n-1}, x_n)x_n$$

$$= l_{t_1}(x_1)q_{t_2-t_1}(x_1, x_2) \cdots q_{t_n-t_{n-1}}(x_{n-1}, x_n). \qquad (6.5.4)$$

There is the relationship:

$$p_t^{3\uparrow}(x, y) = \frac{q_t(x, y)}{xy}$$

as well as

$$\frac{q_t(x, y)}{x} \underset{x\to 0}{\longrightarrow} l_t(y)$$

which is seen from (6.5.2). Hence,

$$p_{t_1}^{3\uparrow}(0, x_1) = \frac{l_{t_1}(x_1)}{x_1}$$

and the relation (6.5.4) now follows easily.

Remark 6.5.2. We may avoid the limiting procedure (6.5.1) in order to prove (6.4.4), by using directly the formula:

$$\Pi^r(\Gamma_{t_n}) = E_0^{(3)}\left[\Gamma_{t_n}\left(\frac{r}{r-t_n}\right)^{3/2}\exp\left(-\frac{X_{t_n}^2}{2(r-t_n)}\right)\right]$$

(see Sect. 1.5), with

$$\Gamma_{t_n} = \mathbf{1}_{(X_{t_1}\in A_1,\cdots,X_{t_n}\in A_n)}.$$

6.6   Proof of the Markovianity of n

Theorem 6.6.1 (Reminder: The Markov property for n). *Under* **n**, *the process* $(\varepsilon(t); t > 0)$ *is Markovian, and the entrance law is:*

$$\lambda_t(dy) = \mathbf{n}_t(dy) = l_t(|y|)dy,$$

where

$$l_t(y) = \frac{1}{\sqrt{2\pi t^3}}|y|e^{-y^2/2t},$$

and, by the symmetry principle, the semigroup is:

$$Q_t(x;dy) = \mathbf{W}_x(X_t \in dy; t < T_0)$$
$$\equiv \frac{1}{\sqrt{2\pi t}}\left(\exp\left(-\frac{1}{2t}(y-x)^2\right) - \exp\left(-\frac{1}{2t}(x+y)^2\right)\right)dy.$$

We shall now consider the Poisson point process with Itô measure **n** to show that:

$$\mathbf{n}\left(\big(\varepsilon(r) \in A\big)\mathbf{1}_\Gamma \circ \theta_r\right) = \mathbf{n}\left(\big(\varepsilon(r) \in A\big)Q_{\varepsilon(r)}(\Gamma)\right). \qquad (6.6.1)$$

We relativize the process $(e_l; l > 0)$ to $\{\varepsilon : \varepsilon(r) \in A\}$; for example, we can choose: $A \subset]0, \infty[$.

We utilize the following lemma:

Lemma 6.6.2. *Let m be the characteristic measure of a PPP, such that* $m(U) < \infty$ *and* $\Gamma \subset U$, *and let* $S = \inf\{t : N_t^U > 0\}$; *it is an exponential variable with parameter* $m(U)$. *Then, S and* e_S *are independent and*

$$P(e_S \in \Gamma) = \frac{m(\Gamma)}{m(U)}.$$

Proof. We know that (N_t^Γ) and $(N_t^{\Gamma^c})$ are independent, with parameters $m(\Gamma)$ and $m(\Gamma^c)$, respectively; and let T and T' be the first jumping times for (N_t^Γ) and $(N_t^{\Gamma^c})$, respectively. Then:

$$\{S > t; e_S \in \Gamma\} \equiv \{t < T < T'\}.$$

Hence:

$$
\begin{aligned}
P(S > t; e_S \in \Gamma) &= \int_t^\infty m(\Gamma)e^{-m(\Gamma)u}du \int_u^\infty m(\Gamma^c)e^{-m(\Gamma^c)v}dv \\
&= \int_t^\infty m(\Gamma)e^{-m(\Gamma)u}du \int_0^\infty dh\, m(\Gamma^c)e^{-m(\Gamma^c)(u+h)} \quad (v = u + h) \\
&= \int_t^\infty m(\Gamma)e^{-m(U)u}du \\
&= \left(\frac{m(\Gamma)}{m(U)}\right) m(U) \int_t^\infty e^{-m(U)u}du \\
&= \left(\frac{m(\Gamma)}{m(U)}\right) P(S > t).
\end{aligned}
$$

$\square$

Now, we shall focus more precisely on the Markov property, by proving (6.6.1).

Proof of (6.6.1). We discuss in relation to the Poisson point process: $(e_l^{\{\varepsilon(r)\in A\}}$; $l > 0)$. The equalities (6.6.1) and (6.6.2) below are equivalent, thanks to Lemma 6.6.2.

$$P\left(e_.^{\{\varepsilon(r)\in A\}} \in \theta_r^{-1}(\Gamma)\right) = \frac{\mathbf{n}\left(1_A(\varepsilon_r)Q_{\varepsilon(r)}(\Gamma)\right)}{\mathbf{n}\left(1_A(\varepsilon_r)\right)}. \tag{6.6.2}$$

Let S be the first jumping time of $e_.^{\{\varepsilon(r)\in A\}}$, where $\{\varepsilon(r) \in A\} \equiv U$ as of Lemma 6.6.2; accordingly, τ_S and τ_{S-} are $(\mathcal{F}_t)$ stopping times, and $T = \tau_{S-} + r$.

The LHS of (6.6.2) is equal to:

$$
\begin{aligned}
(6.6.2) &= P\left((B_T \in A) \cap (\hat{B} \circ \theta_T \in \Gamma)\right) \\
&= P\left((B_T \in A)Q_{B_T}^{\hat{W}}(\Gamma)\right) \\
&= \int_A \gamma(dx)Q_x^{\hat{W}}(\Gamma),
\end{aligned}
$$

where $Q_x^{\hat{W}}$ is the law of $\hat{B}$, Brownian motion killed at 0, and γ is the law of B_T. Indeed, from the previous lemma:

$$\gamma(dx) = \frac{\mathbf{n}(\varepsilon(r) \in dx)}{\mathbf{n}(\mathbf{1}_A(\varepsilon(r)))} \tag{6.6.3}$$

and (6.6.2) has been proven.   □

Now, it remains to prove the "full" Markov property. We consider:

$$\mathbf{n}\left(\left(\prod_{i=1}^{k} f_i\big(\varepsilon(t_i)\big)\right)\mathbf{1}_\Gamma \circ \theta_{t_k}\right)$$

$$= \mathbf{n}\left(f_1\big(\varepsilon(t_1)\big) F \circ \theta_{t_1}\right),$$

$$\text{where } F = \prod_{i=2}^{k} f_i\big(\varepsilon(t_i - t_1)\big)\mathbf{1}_\Gamma \circ \theta_{t_k - t_1}$$

$$= \mathbf{n}\left(f_1(\varepsilon(t_1)) Q_{\varepsilon(t_1)}(F)\right)$$

$$= \mathbf{n}\left(f_1(\varepsilon(t_1)) Q_{\varepsilon(t_1)}\left(\prod_{i=2}^{k} f_i\big(\varepsilon(t_i - t_1)\big) Q_{\varepsilon(t_k - t_1)}(\Gamma)\right)\right)$$

$$= \mathbf{n}\left(\prod_{i=1}^{k} f_i(\varepsilon(t_i)) Q_{\varepsilon(t_k)}(\Gamma)\right).$$

We have now demonstrated the Markov property completely.

6.7 The Formula for Entrance Laws

(a) We shall use the additive formula to prove the formula for the entrance laws.

$$E_{x=0}\left[\int_0^\infty dt\, e^{-at} f(X_t)\right] = E\left[\sum_\lambda \int_{\tau_{\lambda-}}^{\tau_\lambda} dt\, e^{-at} f(X_t)\right]$$

$$= E\left[\int_0^\infty d\lambda\, e^{-a\tau_\lambda-} \int \mathbf{n}(d\varepsilon) \int_0^{V(\varepsilon)} e^{-at} f(\varepsilon_t) dt\right]$$

$$= E\left[\int_0^\infty d\lambda\, e^{-\lambda\sqrt{2a}} \int \mathbf{n}(d\varepsilon) \int_0^{V(\varepsilon)} e^{-at} f(\varepsilon_t) dt\right]$$

and on the other hand, the left side of the above equation is equal to:

$$E_{x=0}\left[\int_0^\infty dt\, e^{-at} f(X_t)\right]$$

$$= \int_0^\infty dt\, e^{-at} \int_{-\infty}^\infty dy\, f(y)\frac{e^{-y^2/2t}}{\sqrt{2\pi t}}$$

$$= \frac{1}{\sqrt{2a}}\int_{-\infty}^\infty dy\, f(y)e^{-\sqrt{2a}|y|}, \qquad (6.7.1)$$

since

$$\int_0^\infty dt\, e^{-at}\frac{1}{\sqrt{2\pi t}}e^{-y^2/2t} = \frac{1}{\sqrt{2a}}e^{-\sqrt{2a}|y|}. \qquad (6.7.2)$$

(b) To complete the proof, going back to (6.7.1), we have obtained:

$$\int_{-\infty}^\infty dy f(y)e^{-\sqrt{2a}|y|} = \int \mathbf{n}(d\varepsilon)\int_0^{V(\varepsilon)} e^{-at} f(\varepsilon_t)dt$$

$$= \int dt\, e^{-at} l_t(f).$$

The LHS equals:

$$\int_{-\infty}^\infty dy f(y)\int_0^\infty P(T_y \in dt)e^{-at}.$$

Hence, we obtain the desired formula for $l_t(f)$ by injectivity of the Laplace transform.

6.8   A (Partial) Proof of Williams' Representation of n

(a) The proof relies on Williams' decomposition of the excursion straddling time T_a. We first state this decomposition by means of a figure (Fig. 6.1).

(i) Brownian motion $(B_{g_{T_a}+u}; u \le T_a - g_{T_a})$ is a BES(3) considered up to its first hitting time of a;

(ii) $(B_{T_a+u}; u \le d_{T_a} - T_a)$ is a Brownian motion starting from a, and considered up to its first hitting time of 0. Its maximum M_a is distributed as a/U (see Sect. 1.8).

Now comes the main statement:

conditionally on $M_a = m$, both processes $(B_{T_a+u}; u \le \rho_{M_a} - T_a)$ and $(B_{d_{T_a}-u}; u \le d_{T_a} - \rho_{M_a})$ are two independent BES(3) processes considered up to their first hitting time of m.

Fig. 6.1 The excursion straddling T_a

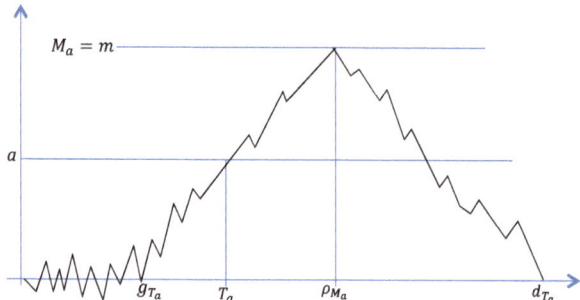

(iii) Putting (i) and (ii) together, conditionally on $M_a = m$, the processes: $(B_{g_{T_a}+u}; u \leq \rho_{M_a})$ and $(B_{d_{T_a}-u}; u \leq d_{T_a}-\rho_{M_a})$ are two iid BES(3) processes considered up to their first hitting time of m.

(b) We shall not prove Williams' decomposition completely, only part (i). See below.

(c) For now, we prove how this Williams decomposition entails Williams' representation of **n**.

Consider, on Ω_*, the set $U_a = \{\varepsilon : M(\varepsilon) > a\} \equiv \{\varepsilon : V(\varepsilon) > T_a\}$. We then obtain, from the second isolation formula (7.5.3) below:

$$\mathbf{n}_+(\Gamma \cap U_a) = \mathbf{n}_+(U_a)\mathbf{W}(e^a \in \Gamma) \qquad (6.8.1)$$

where e^a denotes the excursion straddling time T_a.

We know, from Sect. 6.1, that: $\mathbf{n}_+(U_a) = 1/2a$, so that (6.8.1) becomes:

$$\mathbf{n}_+(\Gamma \cap U_a) = \frac{1}{2a}\mathbf{W}(e^a \in \Gamma). \qquad (6.8.1')$$

Now, from Williams' decomposition,

$$\mathbf{W}(e^a \in \Gamma) = a \int_a^\infty \frac{dx}{x^2} \mathbf{N}(x, \Gamma) \qquad (6.8.2)$$

where $\mathbf{N}(x, \Gamma)$ is the kernel obtained by putting two independent BES(3) processes back to back until they reach level x.

Consider jointly (6.8.1') and (6.8.2), we obtain:

$$\mathbf{n}_+(\Gamma) = \frac{1}{2} \int_0^\infty \frac{dx}{x^2} \mathbf{N}(x, \Gamma)$$

or, more completely, for $\varphi : \mathbb{R}_+ \to \mathbb{R}_+$, and F any ≥ 0 functional on Ω_*:

$$\mathbf{n}_+(F\varphi(M)) = \frac{1}{2} \int_0^\infty \frac{dx}{x^2} \, \varphi(x) \int \mathbf{N}(x, d\eta) \, F(\eta)$$

which is nothing else but Williams' representation of $\mathbf{n}_+$, as presented in Sect. 6.1.

(d) As said in (b), we only prove (i) of Williams' decomposition in (a).

It is a straightforward consequence of the enlargement formula of the Brownian filtration in order to make g_{T_a} a stopping time that: $(B_{g_{T_a}+u}; u \leq T_a - g_{T_a})$ is a BES(3) process; it suffices to apply the formula in Sect. 1.9 after computing

$$Z_t^{g_{T_a}} = 1 - \frac{B_t^+}{a}, \quad t < T_a$$

and one finds that $(B_{g_{T_a}+u}; u \leq T_a - g_{T_a})$ satisfies the SDE of BES(3), namely:

$$B_{g_{T_a}+u} = \tilde{\beta}_u + \int_0^u \frac{ds}{B_{g_{T_a}+s}}, \quad u \leq T_a - g_{T_a}.$$

References

1. L.C.G. Rogers, Williams' characterisation of the Brownian excursion law: proof and applications. Seminar on Probability, XV (Univ. Strasbourg, Strasbourg, 1979/1980) (French). *Lecture Notes in Math.*, vol. 850. (Springer, Berlin, 1981), pp. 227–250
2. D. Revuz, M. Yor, Continuous martingales and Brownian motion. *Grundlehren der Mathematischen Wissenschaften [Fundamental Principles of Mathematical Sciences]*, vol. 293, 3rd edn. (Springer, Berlin, 1999)

Chapter 7
A Simple Path Decomposition of Brownian Motion Around Time $t = 1$

Random Brownian scaling allows to represent the Brownian bridge, the normalized Brownian excursion, and the Brownian meander and co-meander.

Properties of the meander are used to study Azéma martingale.

Notation: Given a process $(X_t; t \geq 0)$, we shall denote

$$X^{[a,b]} = \left(\frac{1}{\sqrt{b-a}} X_{a+t(b-a)}; t \leq 1 \right).$$

This defines the operation of Brownian scaling over the time interval $[a, b]$ (Fig. 7.1).

7.1 Another Representation of the Brownian Bridge

Besides the easily proven Brownian bridge representation $(B_u - uB_1, u \leq 1)$, we show the Brownian scaling representation:

Theorem 7.1.1. *Given a BM $(B_u; u \geq 0)$, the process $B^{[0,g_1]}$ is a Brownian bridge independent of $\sigma\{g_1, B_{g_1+u}; u \geq 0\}$.*

Proof. By time inversion, $B_t = t \hat{B}_{1/t}$ where $\hat{B}$ is a BM. Note that $\hat{d}_1 = \inf\{u > 1 : \hat{B}_u = 0\} = \frac{1}{g_1}$; then

$$\frac{1}{\sqrt{g_1}} B_{tg_1} = t \sqrt{g_1} \hat{B}_{\frac{1}{tg_1}} = \frac{t}{\sqrt{\hat{d}_1}} \left(\hat{B}_{\frac{1}{g_1}+\frac{1}{g_1}(\frac{1}{t}-1)} - \hat{B}_{\frac{1}{g_1}} \right) = \frac{t}{\sqrt{\hat{d}_1}} \left(\hat{B}_{\hat{d}_1+\hat{d}_1(\frac{1}{t}-1)} - \hat{B}_{\hat{d}_1} \right).$$

Since $(\hat{B}_{\hat{d}_1+u} - \hat{B}_{\hat{d}_1}; u \geq 0)$ is a BM independent of $\hat{\mathcal{F}}_{\hat{d}_1}$ and $\hat{B}_{\hat{d}_1} = 0$, the process $\tilde{B}_u = \frac{1}{\sqrt{\hat{d}_1}} \hat{B}_{\hat{d}_1+\hat{d}_1 u}$ is also a BM independent of $\hat{\mathcal{F}}_{\hat{d}_1}$; hence $t \tilde{B}(\frac{1}{t}-1)$ is a Brownian bridge independent of $\hat{\mathcal{F}}_{\hat{d}_1}$; this implies the theorem. $\qquad\square$

J.-Y. Yen and M. Yor, *Local Times and Excursion Theory for Brownian Motion,*
Lecture Notes in Mathematics 2088, DOI 10.1007/978-3-319-01270-4_7,
© Springer International Publishing Switzerland 2013

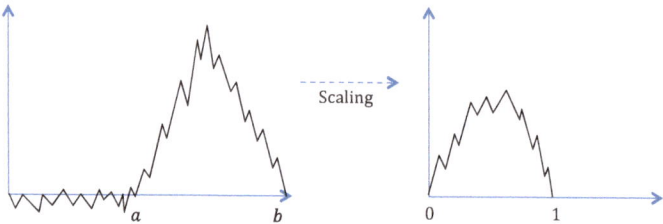

Fig. 7.1 Brownian scaling operation

7.2 The Normalized Brownian Excursion

Theorem 7.2.1. $|B|^{[g_1,d_1]}$ *is distributed as BES(3) bridge (abbreviated BBES(3))*
and is independent of

$$\hat{\mathcal{F}}_{g_1],[d_1} = \sigma\{B_u; u \leq g_1\} \vee \sigma\{B_u; u \geq d_1\} \vee \sigma\{sgn(B_1)\}.$$

The proof is given later in Sect. 7.5 (Fig. 7.2).
Remark (and Exercise!). Our task would be extremely simplified if one could
construct a proof similar to that for the Brownian bridge in Sect. 7.1.
Note that:

$$r(u) \equiv \frac{1}{\sqrt{d_1 - g_1}} B_{[g_1 + u(d_1 - g_1)]}$$

$$\equiv \frac{1}{\sqrt{d_1 - g_1}} [g_1 + u(d_1 - g_1)] \hat{B}_{\frac{1}{g_1 + u(d_1 - g_1)}}$$

But, $d_1 = 1/\hat{g}_1; g_1 = 1/\hat{d}_1$. Therefore:

$$r(u) = \frac{1}{\sqrt{\frac{1}{\hat{g}_1} - \frac{1}{\hat{d}_1}}} \hat{\theta}(u) \hat{B}_{\frac{1}{\hat{\theta}(u)}},$$

where $\hat{\theta}(u) = \frac{1}{\hat{d}_1} + u\left(\frac{1}{\hat{g}_1} - \frac{1}{\hat{d}_1}\right)$.

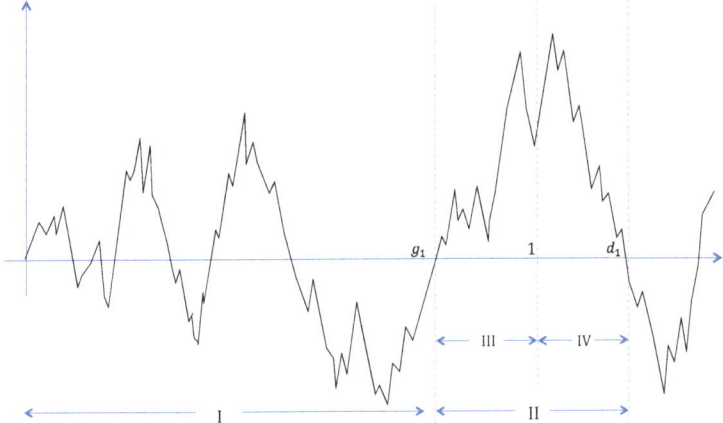

Fig. 7.2 Path decomposition up to d_1.

Therefore, we get:

$$r(u) = \Phi_u(\hat{g}_1, \hat{d}_1)\,\hat{B}_{1/\hat{\theta}(u)}.$$

$$\Phi_u(\hat{g}_1, \hat{d}_1) = \frac{\sqrt{\hat{g}_1 \hat{d}_1}}{\sqrt{\hat{d}_1 - \hat{g}_1}}\left(\frac{\hat{g}_1 + u(\hat{d}_1 - \hat{g}_1)}{\hat{g}_1 \hat{d}_1}\right)$$

$$= \frac{\hat{g}_1 + u(\hat{d}_1 - \hat{g}_1)}{\sqrt{\hat{g}_1 \hat{d}_1}\,\sqrt{\hat{d}_1 - \hat{g}_1}}\,\hat{B}_{1/\hat{\theta}(u)}.$$

Thus, the time inversion trick does no longer seem to lead anywhere!

7.3   The Brownian Meander

Theorem 7.3.1 (Imhof). *Let* **M** *denote the distribution of the Brownian meander* $m \equiv |B|^{[g_1,1]}$. *Then*

$$\mathbf{M} = \sqrt{\frac{\pi}{2}}\,\frac{1}{X_1}\,P_0^{(3)}\big|_{\mathcal{F}_1}$$

where $P_0^{(3)}$ *is the law of the* $BES_0(3)$ *process.*

Proof. We first write:

$$\frac{1}{\sqrt{1-g_1}}|B_{g_1+u(1-g_1)}| = \frac{\sqrt{d_1-g_1}}{\sqrt{1-g_1}}\frac{1}{\sqrt{d_1-g_1}}\left|B_{g_1+u\left(\frac{1-g_1}{d_1-g_1}\right)(d_1-g_1)}\right|$$

$$\equiv \frac{1}{\sqrt{T}}r(uT),$$

where $T = \frac{1-g_1}{d_1-g_1}$ and $r \equiv |B|^{[g_1,d_1]}$. We now prove that T has density $\frac{1}{2\sqrt{t}}$ on $[0,1]$. Indeed, the joint law of (g_1, d_1) is easily seen to be

$$P(g_1 \in du, d_1 \in dt) = \frac{du\,dt}{2\pi\,\sqrt{u}(t-u)^{3/2}}, \quad (u < 1; t > 1)$$

from which one deduces that

$$T = \frac{1-g_1}{(d_1-1)+(1-g_1)}$$

has the desired density.

Moreover, by Theorem 7.2.1, T is independent of r. Thus:

$$E\left[F\left(\frac{1}{\sqrt{T}}r(uT); u \le 1\right)\right] = \int_0^1 \frac{ds}{2\sqrt{s}}E\left[F\left(\frac{1}{\sqrt{s}}r(us); u \le 1\right)\right]$$

$$= \int_0^1 \frac{ds}{2\sqrt{s}}E_0^{(3)}\left[F\left(\frac{X_{us}}{\sqrt{s}}; u \le 1\right)\frac{1}{(1-s)^{3/2}}e^{-X_s^2/2(1-s)}\right] \quad \text{(from Sect. 1.5)}$$

$$= \int_0^1 \frac{ds}{2\sqrt{s}}E_0^{(3)}\left[F\left(X_u; u \le 1\right)\frac{1}{(1-s)^{3/2}}e^{-sX_1^2/2(1-s)}\right]$$

$$= E_0^{(3)}\left[F(X_u; u \le 1)\int_0^1 \frac{ds}{2s^{1/2}(1-s)^{3/2}}e^{-sX_1^2/2(1-s)}\right].$$

Using the change of variables $\frac{s}{1-s} = u$, we find that the above expectation is equal to

$$E_0^{(3)}\left[F(X_u; u \le 1)\sqrt{\frac{\pi}{2}\frac{1}{X_1}}\right].$$

$\square$

There is a close relationship between $\mathbf{n}$ and the law of the meander.

Proposition 7.3.2. *For any variable F_t which is $\mathcal{F}_t$-measurable and positive*

$$\mathbf{n}_+(F_t \mathbf{1}_{(t<V)}) = E_0^{(3)}\left(F_t\frac{1}{2X_t}\right).$$

Proof. We use the decomposition with respect to V.

$$\mathbf{n}_+(F_t\mathbf{1}_{(t<V)}) = \int_t^\infty \frac{dv}{2\sqrt{2\pi v^3}} \Pi^v(F_t),$$

where Π^v denotes the law of a BES(3) bridge of length v. Thus, using Sect. 1.5, we obtain, for $t < v$:

$$\Pi^v(F_t) = E_0^{(3)}\Big[F_t\Big(\frac{v}{v-t}\Big)^{3/2} \exp\Big(-\frac{X_t^2}{2(v-t)}\Big)\Big].$$

Using Fubini's theorem,

$$\mathbf{n}_+(F_t\mathbf{1}_{(t<V)}) = \frac{1}{2\sqrt{2\pi}}E_0^{(3)}\Big[F_t\int_0^\infty \frac{du}{u^{3/2}}\exp\Big(-\frac{X_t^2}{2u}\Big)\Big]$$

$$= E_0^{(3)}\Big(F_t\frac{1}{2X_t}\Big).$$

□

Remark 7.3.3 (Informal). Note that from Theorem 7.3.1, we have:

$$\mathbf{n}_+(F_1\mathbf{1}_{(1<V)}) = E_0^{(3)}\Big(F_1\frac{1}{2X_1}\Big)$$

$$= c\,\mathbf{M}(F_1),$$

for a universal constant c. Therefore,

$$\mathbf{n}_+(F_1|\varepsilon(1) = 0, V \geq 1) = P_0^{(3)}(F_1|X_1 = 0),$$

from which we should be able to deduce

$$\mathbf{n}_+(F_1|V = 1) = \Pi^{(1)}(F_1)$$

our desired BES bridge! More generally, the same "argument" leads to:

$$\mathbf{n}_+(F_t|V = t) = \Pi^{(t)}(F_t).$$

7.4   The Brownian Co-meander

Theorem 7.4.1. *The co-meander $\tilde{m} \equiv \Big(\frac{1}{\sqrt{d_1-1}}|B_{d_1-u(d_1-1)}|; u \leq 1\Big)$ has law*

$$\tilde{\mathbf{M}} = \frac{1}{X_1^2}P_0^{(3)}|_{\mathcal{F}_1}.$$

Proof. Define:

$$\tilde{T} = \frac{d_1 - 1}{d_1 - g_1} = 1 - T, \quad \tilde{r}(u) = \frac{1}{\sqrt{d_1 - g_1}} |B_{d_1 - u(d_1 - g_1)}| = r(1 - u).$$

From the stability by time reversal of the bridge's law, we get:

$$E\left[F\left(\frac{1}{\sqrt{\tilde{T}}} \tilde{r}(u\tilde{T}); u \le 1\right)\right]$$

$$= \int_0^1 \frac{ds}{2\sqrt{1-s}} E\left[F\left(\frac{1}{\sqrt{s}} r(1 - us); u \le 1\right)\right]$$

$$= \int_0^1 \frac{ds}{2\sqrt{1-s}} E\left[F\left(\frac{1}{\sqrt{s}} r(us); u \le 1\right)\right]$$

$$= \int_0^1 \frac{ds}{2(1-s)^2} E_0^{(3)}\left[F\left(\frac{X_{us}}{\sqrt{s}}; u \le 1\right) e^{-X_s^2/2(1-s)}\right]$$

$$= \int_0^1 \frac{ds}{2(1-s)^2} E_0^{(3)}\left[F\left(X_u; u \le 1\right) e^{-sX_1^2/2(1-s)}\right]$$

$$= E_0^{(3)}\left[F(X_u; u \le 1) \int_0^\infty \frac{dv}{2} \exp\left(-\frac{vX_1^2}{2}\right)\right]$$

$$= E_0^{(3)}\left[F(X_u; u \le 1) \frac{1}{X_1^2}\right].$$

□

Theorem 7.4.2. *(1) The process $\tilde{m}$ is independent of $\sigma\{B_u; u \le 1\} \vee \sigma\{B_u; u \ge d_1\}$.*
(2) $\tilde{m}(1) \equiv \frac{|B_1|}{\sqrt{d_1 - 1}}$ has the same distribution as $|B_1|$ and is independent of B_1.
(3) $\tilde{m}$ has the same distribution as

$$R^{[0, \gamma^1]} \equiv \left(\frac{1}{\sqrt{\gamma^1}} R_{u\gamma^1}; u \le 1\right)$$

where $\gamma^1 \equiv \sup\{v \ge 0 : R_v = 1\}$, and R denotes a $BES_0(3)$ process.

Proof. (1) $\tilde{m}$ is measurable with respect to $\sigma\{r, T\}$ (with the notation in the proof of Theorem 7.4.1), and r, T and the σ-field $\tilde{\mathcal{F}}_{g_1],[d_1}$ introduced in Theorem 7.2.1, are independent.

(2) We have already remarked several times that, conditionally on $\mathcal{F}_1$, the random variable $d_1 - 1$ is distributed as $B_1^2 \hat{T}_1$, which yields the desired result.

(3) Write: $\tilde{m}(u) = \frac{1}{\sqrt{d_1-1}}|B_{1+(d_1-1)-u(d_1-1)}|$; since $d_1 = 1 + \hat{T}_{(-B_1)}$,

$$\tilde{m}(u) = \frac{1}{\sqrt{\hat{T}_{(-B_1)}}}|B_{1+(1-u)\hat{T}_{(-B_1)}} - B_1 + B_1|$$

$$\overset{\text{(law)}}{=} \frac{1}{|B_1|\sqrt{\hat{T}_1}}|-B_1\hat{B}_{(1-u)\hat{T}_1} + B_1| = \frac{1}{\sqrt{\hat{T}_1}}|-\hat{B}_{(1-u)\hat{T}_1} + 1|$$

$$\overset{\text{(law)}}{=} \frac{1}{\sqrt{\gamma^1}}R_{u\gamma^1}, \quad \text{(from Williams' time reversal result, see Sect. 1.6).}$$

$\square$

7.5   Two Isolation Formulae

Although Itô's theorem on Brownian excursions (see Sect. 5.2) is mainly a global description of all excursions as a PPP (Poisson Point Process) it is possible to isolate a given excursion, i.e. an excursion straddling a random time T, and to express its law in terms of $\mathbf{n}$ (Fig. 7.3).

In fact, we shall consider essentially two types of time: $T = t$, a deterministic time, and $T = T_a = \inf\{t : B_t = a\}$, a so-called "terminal time".

For this purpose, we first introduce the notation i_γ, where γ is the left extremity of an excursion. We define the applications: $i_0, i_s : \mathcal{W} = C(\mathbb{R}_+, \mathbb{R}) \to \Omega_* =$ (space of absorbed excursions).

$$\begin{cases} i_0(w) &= w(t), \text{ if } t < V(w); = 0, \text{ after } V(w) \\ i_s(w) &= i_0(\theta_s(w)) \end{cases}$$

where $V(w) = \inf\{u > 0 : w(u) = 0\}$.

This allows to rewrite the *key additive formula* (A) (see Sect. 5.3) in the form

$$E\Big[\sum_{\gamma \in G_w} H(\gamma, w; i_\gamma(w))\Big] = E\Big[\int_0^\infty ds \int H(\tau_s(w), w; \varepsilon)\mathbf{n}(d\varepsilon)\Big]$$

$$= E\Big[\int_0^\infty dL_t \int H(t, w; \varepsilon)\mathbf{n}(d\varepsilon)\Big] \quad \text{(A')}$$

where G_w denotes the set of all left extremities of excursions away from 0, for the trajectory w.

Fig. 7.3 The excursion
straddling t

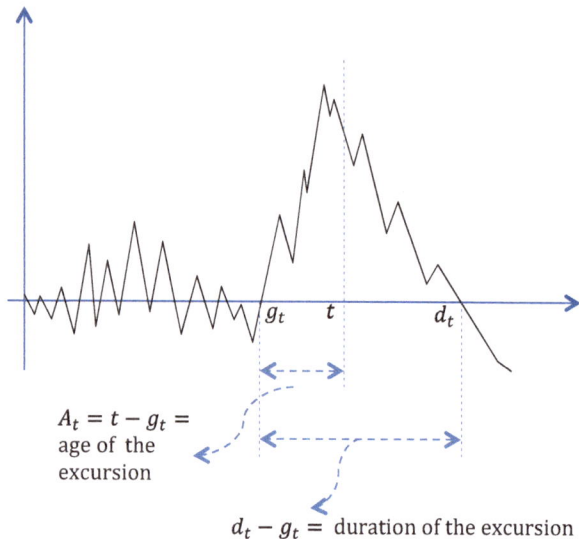

$A_t = t - g_t =$
age of the
excursion

$d_t - g_t =$ duration of the excursion

Proposition 7.5.1. *Let $F \geq 0$ be a generic functional on Ω_*.*

(a) For fixed t, we have:

$$E\big[F(i_{g_t})|\mathcal{F}_{g_t}\big] = q(A_t; F), \qquad (7.5.1)$$

where $q(v, F) = \mathbf{n}(F|V > v)$.

(a') Furthermore, we may refine formula (7.5.1) as:

$$E\big[F(i_{g_t})|\mathcal{F}_{g_t}, \Lambda_t \equiv d_t - g_t\big] = v(\Lambda_t; F), \qquad (7.5.2)$$

where $v(u; F) = \mathbf{n}(F|V = u)$.

(b) For a terminal time T (e.g. $T = T_a$), there is independence of i_{g_T} and $\mathcal{F}_{g_T}$. More precisely, on $0 < g_T < T$, we have:

$$E\big[F(i_{g_T})|\mathcal{F}_{g_T}\big] = \mathbf{n}(F|V > T). \qquad (7.5.3)$$

(b') Furthermore, we may refine formula (7.5.3) as:

$$E\big[F(i_{g_T})|\mathcal{F}_{g_T}, \Lambda_T\big] = \frac{v\big(\Lambda_T; F\mathbf{1}_{(V>T)}\big)}{v\big(\Lambda_T; \mathbf{1}_{(V>T)}\big)}.$$

Sketch of Proof of Proposition 7.5.1.

We shall only give the proof of the identity (7.5.1), the argument for which being at the heart of the "isolation trick" – (For the remaining parts, see Chap. XII of Revuz–Yor [6], Propositions (3.4) and (3.5)).

The isolation trick is that g_t is the only $s \in G_w$ such that $s < t$ and $s + V \circ \theta_s > t$. Thus, if (Z_u) is a positive $(\mathcal{F}_u)$ predictable process, we have:

$$E[Z_{g_t} F(i_{g_t})] = E\Big[\sum_{\gamma \in G_w} Z_\gamma F(i_\gamma) \mathbf{1}_{(V \circ \theta_\gamma > t - \gamma > 0)} \Big].$$

We may now apply the additive formula (A') to obtain:

$$E[Z_{g_t} F(i_{g_t})] = E\Big[\int_0^\infty ds \int Z_{\tau_s}(w) F(\varepsilon) \mathbf{1}_{(V(\varepsilon) > t - \tau_s > 0)} \mathbf{n}(d\varepsilon) \Big]$$

$$= E\Big[\int_0^\infty ds Z_{\tau_s}(w) \mathbf{n}(V > t - \tau_s(w)) q(t - \tau_s(w), F) \Big]$$

and using the additive formula (A') in the reverse direction, this is equal to:

$$E\Big[\sum_{\gamma \in G_w} Z_\gamma q(t - \gamma; F) \mathbf{1}_{(V \circ \theta_\gamma > t - \gamma > 0)} \Big] = E[Z_{g_t} q(t - g_t; F)],$$

which implies (7.5.1).

The arguments for the three other formulae of Proposition 7.5.1 are similar. (We recall that a terminal time satisfies: $T = t + T \circ \theta_t$ on the set $(T > t)$, which we apply with $t = g_T$).

Concerning (a'), i.e: formula (7.5.2), we have already seen, with formula (6.4.1) that $v(u; \cdot)\big(\equiv \mathbf{n}(\cdot | V = u)\big) = \Pi^u$. Hence, we have the following corollaries:

Corollary 7.5.2. *A consequence of the isolation formula (7.5.2) is Sect. 6.1, Statement (1(b)), giving Π^u as the law of $\mathbf{n}_+$ given $V = u$.*

Corollary 7.5.3. *Theorem 7.2.1 follows from Proposition 7.5.1, i.e: the law of the excursion which straddles t, conditioned by $d_t - g_t = u$, is Π^u.*

At this point, it may be worth insisting on the development of our arguments:

- We first proved (formula (6.4.1)) that:

$$\mathbf{n}(\cdot | V = v) = \Pi^v$$

- with formula (7.5.2) and scaling, we have finally obtained that the excursion i_{g_t}, after normalization, is distributed as Π^1, the law of the standard Bessel bridge, which was announced in Theorem 7.2.1.

7.6   Azéma's Martingale and the Brownian Meander

Consider $(\mathcal{F}_t; t \geq 0)$ the Brownian filtration, and two of its subfiltrations, namely $(\mathcal{F}_{g_t}; t \geq 0)$ and $(\tilde{\mathcal{F}}_{g_t}; t \geq 0)$, the latter being defined as

$$\tilde{\mathcal{F}}_{g_t} \equiv \mathcal{F}_{g_t} \vee \{\text{sgn}(B_t)\}.$$

Obviously,

$$\mathcal{F}_{g_t} \subset \tilde{\mathcal{F}}_{g_t} \subset \mathcal{F}_t.$$

Here are some independence properties which will enable us to obtain explicitly the projections of a number of $\mathcal{F}_t$-martingales on the subfiltrations $(\mathcal{F}_{g_t})$ and $(\tilde{\mathcal{F}}_{g_t})$.

Lemma 7.6.1. *For fixed $t > 0$, $\mathcal{F}_{g_t}$, $\text{sgn}(B_t)$ and $m_t(u) \equiv \frac{1}{\sqrt{t - g_t}} |B_{g_t + u(t - g_t)}|$, $u \leq 1$ are independent.*

Proof. It suffices to give the proof for $t = 1$.
From Theorem 7.2.1, $(m_1(u), u \leq 1)$ may be expressed in terms of r, and $\frac{1 - g_1}{d_1 - g_1}$, and this pair is independent from $\tilde{\mathcal{F}}_{g_1}$.
Furthermore, from the balayage formula, we deduce that for every bounded predictable process (k_u), one has:

$$E[k_{g_1} B_1] = 0,$$

thus:

$$E[B_1 | \mathcal{F}_{g_1}] = 0.$$

This conditional expectation equals:

$$E[\text{sgn}(B_1) | \mathcal{F}_{g_1}] \sqrt{1 - g_1} \, E[m_1(1)].$$

Thus, $E[\text{sgn}(B_1) | \mathcal{F}_{g_1}] = 0$, proving the independence of $\text{sgn}(B_1)$ and $\mathcal{F}_{g_1}$, and a fortiori that of $\text{sgn}(B_1)$, $\mathcal{F}_{g_1}$, and m_1.   □

Let $M = (M_t; t \geq 0)$ be a $\mathcal{F}_t$-martingale. Denote

$$\tilde{p}(M)_t = E[M_t | \tilde{\mathcal{F}}_{g_t}] \quad \text{and} \quad p(M)_t = E[M_t | \mathcal{F}_{g_t}].$$

Then, for $c = \sqrt{\frac{\pi}{2}}$, we have

(i) $\tilde{p}(B)_t = c \, \text{sgn}(B_t) \sqrt{t - g_t}, \quad p(B)_t = 0$
(ii) $\tilde{p}(B_t^2 - t)_t = 2(t - g_t) - t = p(B_t^2 - t)_t$
(iii) $\tilde{p}(|B_t| - L_t)_t = c \sqrt{t - g_t} - L_t$

Introduce the notation:

$$\mu_t \equiv \text{sgn}(B_t) \sqrt{t - g_t} \quad \text{and} \quad \nu_t \equiv \sqrt{t - g_t} - \frac{1}{c} L_t.$$

$(\mu_t; t \geq 0)$ and $(\nu_t; t \geq 0)$ are $\tilde{\mathcal{F}}_{g_t}$ martingales. They were defined and studied by Azéma [1] and Azéma–Yor [2]. From (ii), we obtain:

$$\langle \mu \rangle_t = \langle \nu \rangle_t = \frac{t}{2}.$$

Then, the multiple Wiener integrals with respect to μ and ν are well defined, that is,

$$\int_0^\infty d\mu_{s_1} \int_0^{s_1} d\mu_{s_2} \cdots \int_0^{s_{n-1}} d\mu_{s_n} f(s_1, \ldots, s_n)$$

exist for any deterministic Borel function $f \in L^2(\Delta_n)$, where $\Delta_n = \{(s_1, s_2, \ldots, s_n); s_1 > s_2 > \cdots > s_n \geq 0\}$, i.e. such that

$$\int_0^\infty ds_1 \int_0^{s_1} ds_2 \cdots \int_0^{s_{n-1}} ds_n f^2(s_1, \ldots, s_n) < \infty,$$

and the same holds for ν. Denote $\mathcal{M}_\infty = \sigma\{\mu_s; s \geq 0\}$, and $\mathcal{N}_\infty = \sigma\{\nu_s; s \geq 0\}$. The following question arises naturally: Is it possible to develop $L^2(\mathcal{M}_\infty)$ (respectively, $L^2(\mathcal{N}_\infty)$) as a direct sum of the μ-chaoses (respectively, ν-chaoses)? Specifically, we would like to prove

$$L^2(\mathcal{M}_\infty) = \oplus_{n=0}^\infty C_n(\mu),$$

where $C_0(\mu) = \mathbb{R}$, and, for $n \geq 1$:

$$C_n(\mu) = \{X^{(f)} = \int_0^\infty d\mu_{s_1} \int_0^{s_1} d\mu_{s_2} \cdots \int_0^{s_{n-1}} d\mu_{s_n} f(s_1, \ldots, s_n); \ f \in L^2(\Delta_n)\}$$

and similarly for $L^2(\mathcal{N}_\infty)$. The problem for μ has been solved in the affirmative, by M. Émery [5], followed by a series of articles by the same author, whereas, the problem for ν is still unsolved.

Let us prove the result for μ. It is enough to show that every functional of the form

$$X = \prod_{j=1}^n (\mu_{t_j})^{k_j}$$

can be developed in the μ-chaoses for $t_1 < t_2 < \cdots t_n$ and $k_j \in \mathbb{N}$, since the variables of that form are total in $L^2(\mathcal{M}_\infty)$. We shall use a double recurrence argument with the help of which we can diminish both the value of n and of the exponents (k_j). An essential ingredient is the Markov property, which is used in the following argument: We want to calculate $E[X|\tilde{\mathcal{F}}_{g_t}]$, and it is clear that if $t \geq t_n$, $E[X|\tilde{\mathcal{F}}_{g_t}] = X$. On the other hand, if $t_{n-1} < t < t_n$, then

$$E[X|\tilde{\mathcal{F}}_{g_t}] = \prod_{j=1}^{n-1} (\mu_{t_j})^{k_j} E[(\mu_{t_n})^{k_n}|\tilde{\mathcal{F}}_{g_t}].$$

We shall prove the following formulae:

$$E[(\mu_{t'})^{k_n}|\tilde{\mathcal{F}}_{g_t}] = \tilde{P}_{k_n}(\mu_t; t' - t), \quad t < t' \leq t_n, \tag{7.6.1}$$

where $\tilde{P}_{k_n}$ is a polynomial of degree k_n.

$$E[(\mu_{t_n})^{k_n}|\tilde{\mathcal{F}}_{g_t}] = \tilde{P}_{k_n}(\mu_{t_{n-1}}; t_n - t_{n-1}) + \int_{t_{n-1}}^{t} d\mu_s \tilde{Q}_{k_n}(\mu_{s-}, t_n - s), \tag{7.6.2}$$

where

$$\tilde{Q}_{k_n}(x, t) = \frac{\tilde{P}_{k_n}(x, t) - \tilde{P}_{k_n}(0, t)}{x}$$

is a polynomial of degree $k_n - 1$.

Proof. Equation (7.6.2) follows from a general Itô formula for μ which we shall not develop here. Let us prove (7.6.1). Let $s < t$. We shall calculate $E[B_t^k|\tilde{\mathcal{F}}_{g_s}]$ in two different ways.

(a) $E[B_t^k|\tilde{\mathcal{F}}_{g_s}] = E[E[B_t^k|\tilde{\mathcal{F}}_{g_t}]|\tilde{\mathcal{F}}_{g_s}] = E[\mu_t^k c_k|\tilde{\mathcal{F}}_{g_s}],$
    where $c_k = E[m_1^k]$, and we have used the independence between the meander m and $\tilde{\mathcal{F}}_{g_t}$.

(b) $E[B_t^k|\tilde{\mathcal{F}}_{g_s}] = E[E[B_t^k|\mathcal{F}_s]|\tilde{\mathcal{F}}_{g_s}].$
    $E[B_t^k|\mathcal{F}_s]$ can be written in terms of the kth Hermite polynomial, in fact

$$E[B_t^k|\mathcal{F}_s] = H_k(B_s; -(t - s)),$$

where $H_k(x; t) = t^{k/2} H_k\left(\frac{x}{\sqrt{t}}\right)$ may be defined by:

$$\exp\left(\alpha x - \frac{\alpha^2}{2}t\right) = \sum_{k=0}^{\infty} \frac{\alpha^k}{k!} H_k(x; t).$$

Now:

$$E[\exp(\alpha B_t)|\mathcal{F}_s] = \exp(\alpha B_s) \exp\left(\frac{\alpha^2}{2}(t - s)\right),$$

and thus

$$\sum_{k=0}^{\infty} \frac{\alpha^k}{k!} E[B_t^k | \mathcal{F}_s] = \sum_{k=0}^{\infty} \frac{\alpha^k}{k!} H_k(B_s; -(t-s)).$$

It follows that

$$E[B_t^k | \tilde{\mathcal{F}}_{g_s}] = E[H_k(B_s; -(t-s)) | \tilde{\mathcal{F}}_{g_s}] = \tilde{P}_k(\mu_s; -(t-s))$$

and

$$\tilde{P}_k(x;t) = E[H_k(xm_1;t)].$$

Finally, we have proved that

$$E[X | \tilde{\mathcal{F}}_{g_t}] = \begin{cases} X \text{ if } t \geq t_n \\ \prod_{j=1}^{n-1} (\mu_{t_j})^{k_j} E[(\mu_{t_n})^{k_n} | \tilde{\mathcal{F}}_{g_t}], & t_{n-1} < t < t_n, \end{cases}$$

where the last written conditional expectation can be calculated using the recursive formula (7.6.2). The proof finishes with a double recurrence argument.

Notes and Comments.

1. Sections 7.1, 7.2, 7.3, 7.4: Bertoin–Pitman [3] present a unified approach to numerous path transformations connecting the Brownian bridge, excursion and meander. Many important references are found in that paper.
2. Section 7.6: Azéma's martingale $(\mu_t; t \geq 0)$ has some properties which are strikingly similar to those of Brownian motion, e.g.: the chaos representation property (see [2]). Nonetheless, it is very different from Brownian motion, i.e.: it does not have independent increments, it is discontinuous.

   The study of this martingale has been the starting point for the interest (in France, in particular) to the study of the so-called "normal" martingales, i.e.: martingales (M_t) such that $\langle M \rangle_t = t$.

   $\square$

References

1. J. Azéma, Sur les fermés aléatoires. Séminaire de probabilités, XIX, 1983/84. *Lecture Notes in Math.*, vol. 1123. (Springer, Berlin, 1985), pp. 397–495
2. J. Azéma, M. Yor, Étude d'une martingale remarquable. Séminaire de Probabilités, XXIII. *Lecture Notes in Math.*, vol. 1372. (Springer, Berlin, 1989), pp. 88–130
3. J. Bertoin, J. Pitman, Path transformations connecting Brownian bridge, excursion and meander. Bull. des Sci. Math. **118**(2), 147–166 (1994)

4. Ph. Biane, Decompositions of Brownian trajectories and some applications. In *Probability and Statistics; Rencontres Franco-Chinoises en Probabilités et Statistiques; Proceedings of the Wuhan meeting*, ed. by A. Badrikian, P.-A. Meyer, J.-A. Yan (World Scientific, Singapore, 1993), pp. 51–76
5. M. Émery, On the Azéma martingales. Séminaire de Probabilités, XXIII. *Lecture Notes in Math.*, vol. 1372. (Springer, Berlin, 1989), pp. 66–87
6. D. Revuz, M. Yor, Continuous martingales and Brownian motion. *Grundlehren der Mathematischen Wissenschaften [Fundamental Principles of Mathematical Sciences]*, vol. 293, 3rd edn. (Springer, Berlin, 1999)
7. W. Vervaat, A relation between Brownian bridge and Brownian excursion. Ann. Probab. **7**(1), 143–149 (1979)
8. D. Williams, Decomposing the Brownian path. Bull. Am. Math. Soc. **76**, 871–873 (1970)
9. D. Williams, Path decomposition and continuity of local time for one-dimensional diffusions. I. Proc. Lond. Math. Soc. (3) **28**, 738–768 (1974)

Chapter 8
The Laws of, and Conditioning with Respect to, Last Passage Times

The law of a last passage time of a transient diffusion may be expressed in terms of its transition semigroup, while conditioning with respect to a last passage time is related to bridges laws.

8.1 The Bessel Case

The next theorem may be considered as an extension of the previous results about Random Brownian scaling (see Sect. 4.3) to transient diffusions and in particular to the processes $BES(n)$, $n > 2$.

Theorem 8.1.1. *Let* $(R_t; t \geq 0)$ *be a* $BES(n)$ *process,* $n > 2$, *starting from 0, and let*

$$\gamma^a = \sup\{t : R_t = a\}.$$

Then

$$E\left[F\left(\frac{1}{\sqrt{\gamma^a}} R_{u\gamma^a}; u \leq 1\right)\right] = E\left[F\left(R_u; u \leq 1\right)\frac{n-2}{R_1^2}\right] \qquad (8.1.1)$$

$$E\left[F\left(R_u; u \leq \gamma^a\right)|\gamma^a = t\right] = E\left[F\left(R_u; u \leq t\right)|R_t = a\right] \qquad (8.1.2)$$

8.2 General Transient Diffusions

In general, we consider a transient linear diffusion $(X_t; t \geq 0)$, with infinitesimal generator:

$$L = \frac{1}{2}\frac{d}{dm}\frac{d}{ds} \qquad (i)$$

J.-Y. Yen and M. Yor, *Local Times and Excursion Theory for Brownian Motion*,
Lecture Notes in Mathematics 2088, DOI 10.1007/978-3-319-01270-4_8,
© Springer International Publishing Switzerland 2013

where s denotes a scale function such that $s(x) < 0$, $s(\infty) = 0$, and m is the speed measure.

Then, there exists a semigroup density $\dot{p}_t(x, y)$ which is jointly continuous, and symmetric in x and y, such that:

$$(P_t f)(x) = \int m(dy) f(y) \dot{p}_t(x, y).$$

Remark 8.2.1. For simplicity, we also assume that

$$L = \frac{1}{2} a(x) \frac{d^2}{dx^2} + b(x) \frac{d}{dx}, \tag{ii}$$

so that in particular: $\frac{1}{m'(x)} = s'(x) a(x)$.

Throughout this chapter, we keep the conventions (i) and (ii) as taken from Pitman–Yor [6].

If $p_t(x, y)$ denotes the density with respect to dy, we obtain

$$m'(y) \dot{p}_t(x, y) = p_t(x, y) \quad \text{and} \quad \dot{p}_t(x, y) = s'(y) a(y) p_t(x, y).$$

From the occupation density formula for $(s(X_t); t \geq 0)$, we deduce

$$\int_0^t du\ f(X_u) = \int m(dy) f(y) \Lambda_t^{s(y)}$$

where $(\Lambda_t^z; t \geq 0)$ is the local time of $s(X_t)$ at z.

Note how this formula differs from the occupation times formula (2.1.1)

Exercise 8.2.2. Reconcile both formulae by considering:

$$\int_0^t a(X_u) f(X_u) du.$$

Hint: Look for the equality: $L_t^y(X) = \frac{\Lambda_t^{s(y)}}{s'(y)}$.

Theorem 8.2.3. *Let $x < y$.*

The law of $\gamma_y = \sup\{t : X_t = y\}$ is given by $\tag{8.2.1}$

$$P_x(\gamma_y \in dt) = \frac{-1}{2 s(y)} \dot{p}_t(x, y) dt = -\frac{1}{2} \left(\frac{s'(y)}{s(y)} \right) a(y) p_t(x, y) dt.$$

Given $\gamma_y = t$, the process $(X_u; u \leq t)$ has law $P_{x \to y}^{(t)}$ under P_x. $\tag{8.2.2}$

Proof. First we determine the increasing process $(A_t; t \geq 0)$ associated to γ_y. We have, for every previsible process $h \geq 0$:

$$E_x[h_{\gamma_y}] = E_x\left[-\frac{1}{2s(y)} \int_0^\infty h_u d_u(\Lambda_u^{s(y)}) \right] \tag{8.2.3}$$

which implies that the increasing process A, the compensator of γ_y, is:

$$A_t = -\frac{1}{2s(y)} \Lambda_t^{s(y)}; \quad t \geq 0.$$

Indeed, to prove (8.2.3), we remark that:

$$\begin{aligned}
P_x(\gamma_y > t | \mathcal{F}_t) &= P_x\left(\inf_{u \geq t} X_u < y | \mathcal{F}_t\right) \\
&= P_x\left(\sup_{u \geq t}(-s(X_u)) > -s(y) | \mathcal{F}_t\right) \\
&= P_{X_t}\left(\sup_{u \geq 0}(-s(X_u)) > -s(y)\right) \\
&= \frac{s(X_t)}{s(y)} \wedge 1
\end{aligned}$$

where the last two equalities follow from the Markov property and the lemma in Sect. 1.8.

Tanaka's formula implies that

$$\frac{s(X_t)}{s(y)} \wedge 1 = M_t + \frac{1}{2s(y)} \Lambda_t^{s(y)}$$

where M is a martingale and (8.2.3) is then easily obtained by monotone class theorem.

Now, we consider a deterministic function $(\varphi(u); u \geq 0)$ with values in $\mathbb{R}_+$. (8.2.3) implies that:

$$\begin{aligned}
E_x[\varphi(\gamma_y)] &= -\frac{1}{2s(y)} E_x\left[\int_0^\infty \varphi(u) d_u(\Lambda_u^{s(y)}) \right] \\
&= -\frac{1}{2s(y)} \int_0^\infty \varphi(u) d_u \, E_x[\Lambda_u^{s(y)}]. \tag{8.2.4}
\end{aligned}$$

On the other hand,

$$E_x\left[\int_0^t du \, f(X_u) \right] = E_x\left[\int m(dy) f(y) \Lambda_t^{s(y)} \right] = \int m(dy) f(y) E_x[\Lambda_t^{s(y)}].$$

Since we have

$$\int_0^t du\, E_x f(X_u) = \int m(dy) f(y) \int_0^t \dot{p}_u(x,y) du,$$

we obtain:

$$E_x[\Lambda_t^{s(y)}] = \int_0^t du\, \dot{p}_u(x,y)$$

which concludes the proof of (8.2.1).

To prove (8.2.2), we observe that for every bounded previsible process $(h_u; u \geq 0)$, we have:

$$E_x[\varphi(\gamma_y)h_{\gamma_y}] = E_x[\varphi(\gamma_y) E_x[h_{\gamma_y} | \gamma_y]]. \tag{8.2.5}$$

On the other hand,

$$
\begin{aligned}
E_x[\varphi(\gamma_y)h_{\gamma_y}] &= -\frac{1}{2s(y)} E_x\Big[\int_0^\infty \varphi(u)\, E_x[h_u | X_u = y] d_u(\Lambda_u^{s(y)})\Big] \\
&= \int_0^\infty \psi(u)\Big(-\frac{1}{2s(y)}\Big) d_u\, E_x[\Lambda_u^{s(y)}] \\
&= \int_0^\infty \psi(u) P_x(\gamma_y \in du)
\end{aligned}
$$

where

$$\psi(u) = \varphi(u) E_x[h_u | X_u = y],$$

and the last identity holds using the result in (8.2.1), or rather (8.2.4). Then

$$E_x[\varphi(\gamma_y)h_{\gamma_y}] = \int_0^\infty P_x(\gamma_y \in du)\varphi(u) E_x[h_{\gamma_y} | \gamma_y = u].$$

Comparing the above two formulae, we obtain

$$E_x[h_{\gamma_y} | \gamma_y = u] = E_x[h_u | X_u = y]. \qquad \square$$

8.3   Absolute Continuity Relationships up to γ_y

Theorem 8.2.3 will now be our key to relate the laws P_x and Q_x for two transient diffusions on $\mathbb{R}_+$, when restricted to $\mathcal{F}_{\gamma_y}$.

Theorem 8.3.1. *Consider two transient diffusions with laws P_x and Q_x such that*

$$Q_x|_{\mathcal{F}_t} = D_t \cdot P_x|_{\mathcal{F}_t}.$$

Then, the following absolute continuity relation holds:

$$Q_x|_{\mathcal{F}_{\gamma_y}} = (h(y)D_{\gamma_y})P_x|_{\mathcal{F}_{\gamma_y}}, \tag{8.3.1}$$

where $h(y) = \frac{\sigma'(y)}{\sigma(y)}\frac{s(y)}{s'(y)}\frac{\alpha(y)}{a(y)}$; σ and s denote the respective scale functions for Q_x and P_x, and α and a are the respective diffusion coefficients for Q_x and P_x.

Proof. We have

$$P_x[F_{\gamma_y}] = \int_0^\infty dt\, p_t(x, y)\left(-\frac{1}{2}\frac{s'(y)}{s(y)}\right)a(y)P_x[F_t|X_t = y]$$

$$Q_x[F_{\gamma_y}] = \int_0^\infty dt\, q_t(x, y)\left(-\frac{1}{2}\frac{\sigma'(y)}{\sigma(y)}\right)\alpha(y)Q_x[F_t|X_t = y]$$

where we let P_x and Q_x denote as well the expectations with respect to P_x and Q_x, respectively. The theorem then follows from the relation

$$q_t(x, y)Q_x(F_t|X_t = y) = p_t(x, y)P_x(F_t D_t|X_t = y). \qquad \square$$

8.4 Applications

8.4.1 BM with drift considered up to last passage time

Although BM with drift $\mu > 0$ is not a diffusion in $\mathbb{R}_+$, we may apply formally (8.3.1) to BM with different drifts $\mu, \nu > 0$ and corresponding laws $\mathbf{W}^\mu$ and $\mathbf{W}^\nu$. We have:

$$\mathbf{W}^\mu|_{\mathcal{F}_t} = D_t \cdot \mathbf{W}^\nu|_{\mathcal{F}_t}$$

where

$$D_t = \exp\left((\mu - \nu)X_t - (\frac{\mu^2 - \nu^2}{2})t\right).$$

In this case, we have $s_\nu(x) = -\frac{1}{2\nu}\exp(-2\nu x)$ which implies: $\frac{s'_\nu(x)}{s_\nu(x)} = -2\nu$.

Thus, we find:

$$\mathbf{W}^{\mu}|_{\mathcal{F}_{\gamma_a}} = \frac{\mu}{\nu} \exp\left((\mu - \nu)a - (\frac{\mu^2 - \nu^2}{2})\gamma_a\right)\mathbf{W}^{\nu}|_{\mathcal{F}_{\gamma_a}},$$

or, equivalently,

$$\frac{1}{\mu} \exp\left(-\mu a + \frac{\mu^2}{2}\gamma_a\right)\mathbf{W}^{\mu}|_{\mathcal{F}_{\gamma_a}},$$

does not depend on μ. In fact, there exists a σ-finite measure $\mathbf{W}^*$ such that:

$$\mathbf{W}^{\mu}|_{\mathcal{F}_{\gamma_a}} = \mu \exp\left(\mu a - \frac{\mu^2 \gamma_a}{2}\right)\mathbf{W}^*|_{\mathcal{F}_{\gamma_a}}.$$

8.4.2   BES process up to last passage time

Indeed, using time changes, the above discussion may be translated in terms of Bessel processes, which are transient diffusions in $\mathbb{R}_+$. If we call $P^{(\mu)}$ the law of the Bessel process with index μ, or dimension $d = 2(1 + \mu)$, starting from 1 at time 0, we have:

$$P^{(\mu)}|_{\mathcal{F}_t} = R_t^{\mu-\nu} \exp\left(-\frac{(\mu^2 - \nu^2)}{2} \int_0^t \frac{ds}{R_s^2}\right)P^{(\nu)}|_{\mathcal{F}_t}.$$

This may be seen as following from the absolute continuity relation between $\mathbf{W}^{\mu}$ and $\mathbf{W}^{\nu}$ on one hand, and the (Lamperti) time-change relation:

$$\exp(B_t + \mu t) = R^{(\mu)}\left(\int_0^t ds \, \exp(2(B_s + \mu s))\right)$$

on the other hand. Hence, we obtain:

$$P^{(\mu)}|_{\mathcal{F}_{\gamma_r}} = \frac{\mu}{\nu} r^{\mu-\nu} \exp\left(-\frac{\mu^2 - \nu^2}{2} \int_0^{\gamma_r} \frac{ds}{R_s^2}\right)P^{(\nu)}|_{\mathcal{F}_{\gamma_r}},$$

Or, equivalently, there exists a σ-finite measure P^* such that:

$$P^{(\mu)}|_{\mathcal{F}_{\gamma_r}} = \mu r^{\mu} \exp\left(-\frac{\mu^2}{2} \int_0^{\gamma_r} \frac{ds}{R_s^2}\right)P^*|_{\mathcal{F}_{\gamma_r}}.$$

8.4.3   First hit of 0 by Ornstein–Uhlenbeck process

The previous results may be tied together with Lévy's area formula, as the following exercise suggests.

Exercise 8.4.1. Give different proofs of the formula (found in [3]):

$$
^{-\lambda}\mathbf{W}_a(T_0 \in dt) = a\,\frac{\exp\left(\frac{\lambda a^2}{2}\right)}{\sqrt{2\pi}}\,\exp\left(\frac{\lambda}{2}\left(t - a^2\coth(\lambda t)\right)\right)\left(\frac{\lambda}{\sinh(\lambda t)}\right)^{3/2} dt
$$

where $^{-\lambda}\mathbf{W}_a$ denotes the law of a one-dimensional Ornstein–Uhlenbeck process $(U_t; t \geq 0)$, starting from $a \geq 0$, with parameter $(-\lambda)$, $\lambda > 0$, i.e.: (U_t) solves:

$$
dU_t = d\beta_t - \lambda U_t dt; \quad U_0 = a \quad \text{and} \quad T_0 = \inf\{t : U_t = 0\}.
$$

Hint: 1st *proof*:
Use the representation: $U_t = e^{-\lambda t} B_{u_\lambda(t)}$, where $u_\lambda(t) = \frac{e^{2\lambda t}-1}{2\lambda}$, and $(B_u; u \geq 0)$ is a BM.
2nd *proof*:
Use the absolute continuity relation between $^{-\lambda}\mathbf{W}_a$ and $\mathbf{W}_a \equiv {}^0\mathbf{W}_a$, together with D.Williams time reversal result to obtain:

$$
^{-\lambda}\mathbf{W}_a(T_0 > t) = \mathbf{W}_a\left(\mathbf{1}_{(T_0>t)}\exp\left(\frac{\lambda}{2}(a^2 + T_0) - \frac{\lambda^2}{2}\int_0^{T_0} ds\, X_s^2\right)\right)
$$

$$
= E_0^{(3)}\left(\mathbf{1}_{(\gamma_a>t)}\exp\left(\frac{\lambda}{2}(a^2 + \gamma_a) - \frac{\lambda^2}{2}\int_0^{\gamma_a} ds\, X_s^2\right)\right).
$$

Finally, condition with respect to γ_a and use Lévy's area formula (cf. Formula (11.3.1), or indirectly, Theorem 11.5.2).

Notes and Comments.

A detailed discussion of Markovian bridges is presented in Pitman–Yor [6] and in Fitzsimmons–Pitman–Yor [4]. More generally, discussions related to last exits from sets for Markov processes have been extensively studied since Hunt's fundamental memoir (1954); the connections between last exit distributions and capacity theory form an essential part of probabilistic potential theory; see e.g. Chung [1], Getoor–Sharpe [5], Doob's treatise [2] and Stroock's book [8].

References

1. K.L. Chung, Probabilistic approach in potential theory to the equilibrium problem. Ann. Inst. Fourier (Grenoble) **23**(3), 313–322 (1973)
2. J.L. Doob, Classical potential theory and its probabilistic counterpart. *Grundlehren der Mathematischen Wissenschaften [Fundamental Principles of Mathematical Sciences]*, vol. 262. (Springer, New York, 1984)
3. K.D. Elworthy, X.-M. Li, M. Yor, The importance of strictly local martingales; applications to radial Ornstein-Uhlenbeck processes. Probab. Theor Relat. Field **115**(3), 325–355 (1999)
4. P. Fitzsimmons, J. Pitman, M. Yor, Markovian bridges: construction, Palm interpretation, and splicing. *Seminar on Stochastic Processes, 1992*, ed. by R. Bass, **33**, 101–134 (1993)
5. R.K. Getoor, M.J. Sharpe, Last exit times and additive functionals. Ann. Probab. **1**, 550–569 (1973)
6. J. Pitman, M. Yor, Bessel processes and infinitely divisible laws. Stochastic integrals (Proc. Sympos., Univ. Durham, Durham, 1980). *Lecture Notes in Math.*, vol. 851. (Springer, Berlin, 1981), pp. 285–370
7. D. Revuz, M. Yor, Continuous martingales and Brownian motion. *Grundlehren der Mathematischen Wissenschaften [Fundamental Principles of Mathematical Sciences]*, vol. 293, 3rd edn. (Springer, Berlin, 1999)
8. D.W. Stroock, *Probability Theory, An Analytic View* (Cambridge University Press, Cambridge, 1993)

Chapter 9
Integral Representations Relating W and n

In this chapter, we discuss several integral representations between **W** and **n**, based on results in previous chapters. The main result (Theorem 9.1.1) shall play a key role in our derivation of the Feynman–Kac formula in Chap. 10.

9.1 Statement of the Main Theorem

The notation in the following theorem will be explained after its statement.

Theorem 9.1.1. *There is the formula:*

$$\int_0^\infty dt \, \mathbf{W}^t = \left(\int_0^\infty ds \, \mathbf{W}^{\tau_s} \right) \circ \left(\int_{-\infty}^\infty da \; {}^r(\mathbf{W}_a^{T_0}) \right)$$

Notation. Given a random time T, $\mathbf{W}^T$ is the law of $(X_u; u \leq T)$ under **W**, i.e. the Wiener measure defined on the space of continuous functions $(\omega(t); t \leq \zeta_\omega)$ with lifetime $\zeta(\omega)$, thus:

$$\left(\int_0^\infty dt \, \mathbf{W}^t \right) \left(F(X_u; u \leq \zeta) h(\zeta) \right) = \int_0^\infty dt \, h(t) E[F(X_u; u \leq t)];$$

analogously,

$$\left(\int_0^\infty ds \, \mathbf{W}^{\tau_s} \right) \left(F(X_u; u \leq \zeta) h(\zeta) \right) = \int_0^\infty ds \, E[F(X_u; u \leq \tau_s) h(\tau_s)].$$

On the other hand, $\mathbf{W}_a^{T_0}$ denotes the law of a BM starting from a and ending in 0 at T_0, and ${}^r(\mathbf{W}_a^{T_0})$ is the law with reversed time, i.e. ${}^r(\mathbf{W}_a^{T_0}) = (\mathbf{W}_0^{(3)})^{\gamma^a}$ is the law $\mathbf{W}_0^{(3)}$ of a BES(3) process starting from 0 and ending in a at γ^a, its last passage time at a.

J.-Y. Yen and M. Yor, *Local Times and Excursion Theory for Brownian Motion*, 101
Lecture Notes in Mathematics 2088, DOI 10.1007/978-3-319-01270-4_9,
© Springer International Publishing Switzerland 2013

Finally, the symbol $\circ$ stands for the composition of two trajectories, $(\omega(u); u \le \zeta_\omega)$ and $(\omega'(v); v \le \zeta_{\omega'})$ under the condition $\omega(\zeta_\omega) = \omega'(0)$,

$$(\omega \circ \omega') \equiv \begin{cases} \omega(t); \ t \le \zeta_\omega \\ \omega'(t - \zeta_\omega); \ \zeta_\omega \le t \le \zeta_\omega + \zeta_{\omega'} \end{cases}$$

9.2   Proof of the Theorem

To show the theorem, we use the following sequence of identities.

$$\int_0^\infty dt \ \mathbf{W}^t = \left(\int_0^\infty ds \ \mathbf{W}^{\tau_s} \right) \circ \left(\int_0^\infty dt \ \mathbf{n}^t(\cdot; t < V) \right) \qquad (9.2.1)$$

$$\int_0^\infty ds \ \mathbf{W}^{\tau_s} = \int_0^\infty \frac{du}{\sqrt{2\pi u}} Q^u \qquad (9.2.2)$$

where Q^u denotes the law of a Brownian bridge of length u.

$$\int_0^\infty dt \ \mathbf{n}_+^t(\cdot; t < V) = \int_0^\infty da \ (P_0^{(3)})^{\gamma^a} \qquad (9.2.3)$$

$$\int_0^\infty dt \ \mathbf{n}^t(\cdot; t < V) = \int_{-\infty}^\infty da \ ^r(\mathbf{W}_a^{T_0}). \qquad (9.2.4)$$

Equations (9.2.2) and (9.2.3) follow from the absolute continuity results studied in Chap. 4. Equation (9.2.4) follows from (9.2.3) by symmetry and time reversal. We shall prove (9.2.1) and give some details of the proof of (9.2.3).

To prove (9.2.3), consider a test functional $F(\frac{1}{\sqrt{t}} X_{st}; s \le 1) h(t)$. The LHS of (9.2.3) is equal to:

$$\int_0^\infty dt \ h(t) \mathbf{n}_+ \left(F(\frac{1}{\sqrt{t}} X_{st}; s \le 1) \mathbf{1}_{(t<V)} \right)$$

$$= \int_0^\infty dt \ h(t) E_0^{(3)} \left[F(\frac{1}{\sqrt{t}} X_{st}; s \le 1) \frac{1}{2X_t} \right] \quad \text{(from Proposition 7.3.2)}$$

$$= \int_0^\infty \frac{dt \ h(t)}{2\sqrt{t}} E_0^{(3)} \left[F(X_s; s \le 1) \frac{1}{X_1} \right]$$

with the help of the scaling property.

On the other hand, the RHS of (9.2.3) is equal to:

$$\int_0^\infty da \, E_0^{(3)}\left[h(\gamma^a)F\left(\frac{1}{\sqrt{\gamma^a}}X_{u\gamma^a};u \le 1\right)\right]$$

$$= \int_0^\infty da \, E_0^{(3)}\left[h(a^2\gamma^1)F\left(\frac{1}{\sqrt{\gamma^1}}X_{u\gamma^1};u \le 1\right)\right]$$

and by the change of variables $t = a^2\gamma^1$, the above expression is equal to:

$$E_0^{(3)}\left[\int_0^\infty \frac{dt \, h(t)}{2\sqrt{t\gamma^1}}F\left(\frac{1}{\sqrt{\gamma^1}}X_{u\gamma^1};u \le 1\right)\right].$$

According to the result about the law of $X^{[0,\gamma^1]}$ (cf, Sect. 7.4 or Theorem 8.1.1), this quantity is equal to

$$\int_0^\infty \frac{dt \, h(t)}{2\sqrt{t}} E_0^{(3)}\left[F\left(X_u;u \le 1\right)\frac{1}{X_1}\right]$$

from which we obtain (9.2.3).

Exercise 9.2.1. To give another proof of (9.2.3), use Pitman's representation of BES(3) and show that both sides, when applied to an adequate predictable process F_t, are also equal to

$$\frac{1}{2}E_0^{(3)}\left[\int_0^\infty dX_t F_t\right]$$

Hint:

- From Pitman's theorem, $(2J_t - X_t, t \ge 0)$ is a BM where $J_t = \inf_{u \ge t} X_u$.
- The process (J_t) is the inverse of (γ_a).
- There are the identities:

$$E\left[\int_0^\infty da \, F_{\gamma_a}\right] = E\left[\int_0^\infty dJ_t \, F_t\right]$$

$$= \frac{1}{2}E\left[\int_0^\infty dX_t \, F_t\right] = \frac{1}{2}E\left[\int_0^\infty \frac{dt}{X_t} F_t\right]$$

9.3   Proof of (9.2.1)

To show (9.2.1), we use the following product of the test functions

$$F(B_u;u \le t) \equiv F_-(B_u;u \le g_t)F_+(B_{g_t+u};u \le t - g_t).$$

The LHS of (9.2.1) is equal to:

$$E\left[\int_0^\infty dt\ F_-(B_u; u \le g_t)F_+(B_{g_t+u}; u \le t - g_t)\right]$$

$$= E\left[\sum_s F_-(B_u; u \le \tau_s-)\int_{\tau_s-}^{\tau_s} dt\ F_+(B_{\tau_s-+u}; u \le t - \tau_s-)\right].$$

From the additive formula (A), (see Sect. 5.3), the previous quantity is equal to:

$$E\left[\int_0^\infty ds\ F_-(B_u; u \le \tau_s-)\int \mathbf{n}(d\varepsilon)\int_0^{V(\varepsilon)} F_+(\varepsilon_u; u \le v)dv\right]$$

$$= \left(\int_0^\infty ds\ \mathbf{W}^{\tau_s} \circ \int_0^\infty dt\ \mathbf{n}_+(\cdot; t < V)\right)(F_-F_+).$$

Comment: We note that formulae (9.2.1) to (9.2.4) may be extended in the general diffusion framework considered in [5].

References

1. Ph. Biane, Decompositions of Brownian trajectories and some applications. In Probability and Statistics 51–76. *Rencontres Franco-Chinoises en Probabilités et Statistiques*, ed. by A. Badrikian, P.A. Meyer, J.A. Yan. (World Scientific, Singapore, 1993)
2. Ph. Biane, M. Yor, Valeurs principales associées aux temps locaux browniens. Bull. Sci. Math. (2) **111**(1), 23–101 (1987)
3. C. Leuridan, Une démonstration élémentaire d'une identité de Biane et Yor. Séminaire de Probabilités, XXX. *Lecture Notes in Math.*, vol. 1626. (Springer, Berlin, 1996), pp. 255–260
4. J. Pitman, M. Yor, Decomposition at the maximum for excursions and bridges of one-dimensional diffusions. *Itô's Stochastic Calculus and Probability Theory*. (Springer, Tokyo, 1996), pp. 293–310
5. P. Salminen, P. Vallois, M. Yor, On the excursion theory for linear diffusions. Jpn. J. Math. **2**(1), 97–127 (2007)

Part III
Some Applications of Excursion Theory

Chapter 10
The Feynman–Kac Formula
and Excursion Theory

We provide a proof of the Feynman–Kac formula for Brownian motion, using excursion theory up to an independent exponential time.

10.1 Statement of the FK Formula

There are many variants of Feynman–Kac formula (some forms are "weaker" than others). In this section, we shall take as statement for (FK) the following:

Let $q : \mathbb{R} \to \mathbb{R}_+$, $f : \mathbb{R} \to \mathbb{R}_+$ locally bounded. Then, the double Laplace transform satisfies the identity:

$$\int_0^\infty dt \, e^{-\frac{k^2 t}{2}} E\Big[q(B_t)\exp\Big(-\lambda \int_0^t f(B_s)ds\Big)\Big] = \int_{-\infty}^\infty dx \, q(x)U^{\lambda f}(k,x), \quad \text{(FK)}$$

where $U \equiv U^{\lambda f}(k,\cdot)$ is the unique solution of the differential equation:

$$\frac{1}{2}U''(x) = \Big(\frac{k^2}{2} + \lambda f\Big)(x)U(x), \quad x \neq 0,$$

$$U'(x) \text{ exists for } x \neq 0,$$

$$U(x) \underset{|x|\to\infty}{\longrightarrow} 0,$$

$$U'(0+) - U'(0-) = -2.$$

We shall see how excursion theory, and in particular the independence of the pre-g_θ and post-g_θ processes, where θ is an exponential variable with parameter $(k^2/2)$, explain well our statement of (FK), and, in particular the jump formula:

$$U'(0+) - U'(0-) = -2.$$

J.-Y. Yen and M. Yor, *Local Times and Excursion Theory for Brownian Motion*,
Lecture Notes in Mathematics 2088, DOI 10.1007/978-3-319-01270-4_10,
© Springer International Publishing Switzerland 2013

10.2   Proof of FK via Excursion Theory

We note that the double Laplace transform, on the LHS of (FK) may be written as:

$$\left(\frac{2}{k^2}\right) E\left[q(B_\theta) \exp\left(-A_\theta^{\lambda f}\right)\right] \tag{10.2.1}$$

The independence of the pre-g_θ and post-g_θ processes implies in particular that B_θ and L_θ^0 are independent , with:

$$P(B_\theta \in dx) = \frac{k}{2} \exp(-k|x|)dx;$$

$$P(L_\theta^0 \in dl) = k \exp(-kl)dl.$$

Before we proceed, we need to recall the following fact concerning Sturm–Liouville equations:

$$\frac{1}{2}F''(x) = m(x)F(x) \tag{10.2.2}$$

for an unknown function $F : \mathbb{R}_+ \to \mathbb{R}$ where m is a given ≥ 0, and locally integrable Borel function.

There is a unique non-increasing solution of (10.2.2) such that $F(0) = 1$. In the following, we shall consider this solution for

$$m(x) = \frac{k^2}{2} + \lambda f_+, \quad \text{and}$$

$$m(x) = \frac{k^2}{2} + \lambda f_-,$$

where $f_+ = f_{|\mathbb{R}_+}$ and $f_-(x) = f(-x)$, $x \geq 0$. For simplicity, we take $\lambda = 1$ and we denote the respective Sturm–Liouville solutions by $\Phi^{f+}(k, a)$ and $\Phi^{f-}(k, a)$, $a \geq 0$.

Then the obtention of the RHS of (FK) shall be done via a particular application of the following integral representation formulae:

$$E_{\mathrm{W}}\left[F(B_u; 0 \leq u \leq g_\theta)\right] = k \int_0^\infty dl\, E_{\mathrm{W}}\left[\exp\left(-\frac{k^2 \tau_l}{2}\right)F(B_u; u \leq \tau_l)\right] \tag{10.2.3}$$

and

$$E_{\mathrm{W}}\left[G(B_{\theta-u}; u \leq \theta - g_\theta)\right] = \frac{k}{2} \int_{-\infty}^\infty da\, E_a\left[\exp\left(-\frac{k^2}{2}T_0\right)G(B_u; 0 \leq u \leq T_0)\right]. \tag{10.2.4}$$

Formulae (10.2.3) and (10.2.4) follow respectively from formulae (9.2.1) and (9.2.4). In particular, the following formulae hold:

$$E\big[\exp\big(-A_{g_\theta}^f\big)\big] = k \int_0^\infty dl \; E\big[\exp\big(-\big(\frac{k^2}{2}\tau_l + A_{\tau_l}^f\big)\big)\big] \qquad (10.2.5)$$

whereas:

$$E\big[q(B_\theta)\exp\big(-\big(A_\theta^f - A_{g_\theta}^f\big)\big)\big] = \frac{k}{2} \int_{-\infty}^\infty da \; q(a)E_a\big[\exp\big(-\big(\frac{k^2}{2}T_0 + A_{T_0}^f\big)\big)\big]. \qquad (10.2.6)$$

Moreover, on the RHS of (10.2.5) and (10.2.6), we find with e.g. the help of stochastic calculus:

$$E\big[\exp\big(-\big(\frac{k^2}{2}\tau_l + A_{\tau_l}^f\big)\big)\big] = \exp\big(\frac{l}{2}\big[\big(\Phi^{f+}\big)'(k,0+) + \big(\Phi^{f-}\big)'(k,0+)\big]\big) \quad (10.2.7)$$

and

$$E_a\big[\exp\big(-\big(\frac{k^2}{2}T_0 + A_{T_0}^f\big)\big)\big] = \Phi^{f\pm}(k,a) \qquad (10.2.8)$$

(depending on the sign of a), so that finally, multiplying (10.2.5) and (10.2.6) side by side:

$$E\big[q(B_\theta)\exp(-A_\theta^f)\big] = \frac{k^2 \int_0^\infty da\big(q(a)\Phi^{f+}(k,a) + q(-a)\Phi^{f-}(k,a)\big)}{-(\Phi^{f+})'(k,0+) - (\Phi^{f-})'(k,0+)}. \qquad (10.2.9)$$

Comparing (10.2.9) and the RHS of (FK) yields:

$$U^{\lambda f}(k,x) = \frac{2\big(\Phi^{\lambda f+}(k,x)\mathbf{1}_{(x>0)} + \Phi^{\lambda f-}(k,-x)\mathbf{1}_{(x<0)}\big)}{-(\Phi^{\lambda f+})'(k,0+) - (\Phi^{\lambda f-})'(k,0+)}. \qquad (10.2.10)$$

It follows that $U(x) = U^{\lambda f}(k,x)$ satisfies the conditions stated after (FK), and in particular the intriguing $U'(0+) - U'(0-) = -2$.

Note how the presentation of Kac's basic function in (10.2.10) as a ratio reflects the path decomposition at time g_θ.

For a series of applications of formula (FK) via (10.2.10), we refer the reader to Jeanblanc–Pitman–Yor [4].

We shall only give one application, namely: we show how to recover the arc-sine law for the time spent > 0 by Brownian motion, from the simple form of formula (10.2.10) when $f_+(x) = \frac{\lambda_+^2}{2}$, $f_-(x) = \frac{\lambda_-^2}{2}$. Then (FK) formula becomes:

$$E_{\mathbf{W}}\left[\exp\left(-\left(\frac{\lambda_+^2}{2}\int_0^\theta ds\,\mathbf{1}_{(B_s>0)} + \frac{\lambda_-^2}{2}\int_0^\theta ds\,\mathbf{1}_{(B_s<0)}\right)\right)\right] = \frac{k^2}{\nu_+\nu_-},$$

where $\nu_+ = (k^2 + \lambda_+^2)^{1/2}$, $\nu_- = (k^2 + \lambda_-^2)^{1/2}$. Consequently, the random times $\int_0^\theta ds\,\mathbf{1}_{(B_s>0)}$ and $\int_0^\theta ds\,\mathbf{1}_{(B_s<0)}$ are independent with a common gamma distribution with shape parameter $(1/2)$. Lévy's arc-sine law is a well-known consequence.

References

1. M. Kac, On the average of a certain Wiener functional and a related limit theorem in calculus of probability. Trans. Am. Math. Soc. **65**, 401–414 (1946)
2. M. Kac, On the distribution of certain Wiener functionals. Trans. Am. Math. Soc. **65**, 1–13 (1949)
3. M. Kac, On some connections between probability theory and differential and integral equations, in *Proc. Second Berkeley Symp. Math. Stat. Prob.*, ed. by J. Neyman (University of California Press, Berkeley, 1951), pp. 189–215
4. M. Jeanblanc, J. Pitman, M. Yor, The Feynman–Kac formula and decomposition of Brownian paths. Sociedade Brasiliera de Matemática Applicada e Computacional. Matemática Aplicada e Computacional. Comput. Appl. Math. **16**(1), 27–52 (1997)

Chapter 11
Some Identities in Law

Here we present some remarkable identities in law that are obtained using excursion theory, together with some decomposition of the Brownian trajectories. In particular, we shall study the Lévy–Khinchine representation of the BESQ laws Q_x^δ.

11.1 On Linear Combinations of Reflected BM and Its Local Time

Let $k \geq 0$, and consider the process $(k S_t - B_t; t \geq 0)$. In the particular cases $k = 0$, $k = 1$ and $k = 2$, these processes are respectively a BM, a reflected BM and a BES(3) process.

If $k = 1$, the filtration of $(S_t - B_t; t \geq 0)$ is the same as the natural filtration of $(B_t; t \geq 0)$. If $k = 2$, $(2S_t - B_t; t \geq 0)$ has a smaller filtration. For $k \neq 2$, $(k S_t - B_t; t \geq 0)$ has the same filtration as $(B_t; t \geq 0)$ [1].

On the other hand, if $k \neq 0, 1, 2$, $(k S_t - B_t; t \geq 0)$ is not a Markov process. Its occupation measure is, however, very interesting.

Recall that by Lévy's identity:

$$(k S_t - B_t; t \geq 0) \overset{(\text{law})}{=} ((k - 1)L_t + |B_t|; t \geq 0).$$

We have the following:

Theorem 11.1.1. Let $L = L^0(B)$. Then the process $\left(L_\infty^a(|B| + \frac{2}{\delta}L); a \geq 0\right)$ follows the law Q_0^δ, that is the law of a squared Bessel process with dimension δ, starting at 0.

Proof. Let $f : \mathbb{R}_+ \to \mathbb{R}_+$ be a Borel function. Then,

$$E\left[\exp\left(-\int_0^\infty ds\ f\left(|B_s|+\frac{2}{\delta}L_s\right)\right)\right]$$

$$= E\left[\exp\left(-\sum_{l\geq0}\int_{\tau_{l-}}^{\tau_l} ds\ f\left(|B_s|+\frac{2}{\delta}l\right)\right)\right]$$

$$= \exp\left(-\int_0^\infty dl\int_{\Omega_*}\mathbf{n}(d\varepsilon)\left(1-e^{-\int_0^{V(\varepsilon)} ds\ f(|\varepsilon_s|+\frac{2}{\delta}l)}\right)\right),\quad\text{from (M) (see Sect. 5.3).}$$

By the change of variables $\lambda=\frac{2}{\delta}l$, this expression is equal to:

$$\exp\left(-\frac{\delta}{2}\int_0^\infty d\lambda\int_{\Omega_*}\mathbf{n}(d\varepsilon)\left(1-e^{-\int_0^{V(\varepsilon)} ds\ f(|\varepsilon_s|+\lambda)}\right)\right).\qquad(11.1.1)$$

We want to show that the quantity (11.1.1) is equal to

$$Q_0^\delta\left(\exp\left(-\int_0^\infty dx\ f(x)X_x\right)\right).\qquad(11.1.2)$$

If $\delta=2$, Pitman's theorem (Theorem 3.3.1), the second Ray–Knight theorem (Theorem 2.2.4) and Williams' time reversal theorem (Theorem 1.6.1) imply that

$$\left(L_\infty^a(|B|+L);a\geq0\right)\overset{\text{(law)}}{=}\left(L_\infty^a(R);a\geq0\right)\overset{\text{(law)}}{=}\left((R_a^{(2)})^2;a\geq0\right)$$

which gives the desired equality between (11.1.1) and (11.1.2) for this case.

In the general case, the infinite divisibility property

$$Q_0^{(\gamma\delta/2)}\left(\exp\left(-\int_0^\infty dx\ f(x)X_x\right)\right)=Q_0^{(\gamma)}\left(\exp\left(-\int_0^\infty dx\ f(x)X_x\right)\right)^{\delta/2}$$

allows us to reduce the general case to the case $\gamma=2$ by means of the identity

$$Q_0^\delta\left(\exp\left(-\int_0^\infty dx\ f(x)X_x\right)\right)$$

$$= Q_0^{(2)}\left(\exp\left(-\int_0^\infty dx\ f(x)X_x\right)\right)^{\delta/2}$$

$$= \exp\left(-\frac{\delta}{2}\int_0^\infty d\lambda\int\mathbf{n}(d\varepsilon)\left(1-e^{-\int_0^{V(\varepsilon)} ds\ f(|\varepsilon_s|+\lambda)}\right)\right)$$

which implies the equality between (11.1.1) and (11.1.2) for any $\delta>0$.   □

Remark 11.1.2. We see, with the above arguments, how excursion theory helps to relate in general BESQ processes and Brownian local times, thus completing the Ray–Knight theorems (Theorem 2.2.4).

The next theorem gives an explicit Lévy–Khinchine representation of Q_x^δ.

Theorem 11.1.3. *(Lévy–Khinchine representation of Q_x^δ). Consider*

$$Y_\mu = \int_0^\infty \mu(dt)B_t^2,$$

and more generally

$$Y_\mu = \int \mu(dt)X_t,$$

with (X_t) a $BESQ_x(\delta)$ process, for μ a finite measure such that:
$\int_0^\infty t d\mu(t) < \infty$ Then, Y_μ is finite and infinitely divisible. There exist M and N
σ-finite measures on $C(\mathbb{R}_+, \mathbb{R}_+)$, independent of μ, such that:

$$Q_x^\delta\left(\exp\left(-\int d\mu(t)X_t\right)\right) = \exp\left(-\int (xM + \delta N)(d\varphi)\left(1 - e^{-\int d\mu(t)\varphi(t)}\right)\right).$$

The measure M can be obtained from $\mathbf{n}$ in the following way:
M is the image of $\mathbf{n}_+$ by the application $\varepsilon \to L_V^\cdot(\varepsilon) \equiv (L_V^x(\varepsilon), x \geq 0)$.

$$Q_x^0\left(\exp\left(-\int dy\ f(y)X_y\right)\right) = \exp\left(-x\int M(d\varphi)\left(1 - e^{-\int dy f(y)\varphi(y)}\right)\right),$$

since for $x = 1$ and $f \geq 0$ with support in $\mathbb{R}_+$, we have for Q_1^0:

$$E\left[\exp\left(-\int_0^{\tau_1} ds\ f(B_s))\right] = \exp\left(-\int \mathbf{n}_+(d\varepsilon)\left(1 - e^{-\int_0^{V(\varepsilon)} dsf(\varepsilon_s)}\right)\right).$$

Exercise 11.1.4. Express N in terms of $\mathbf{n}$, with the help of Theorem 11.1.1.

Remark 11.1.5. This argument can be extended to square Ornstein–Uhlenbeck processes of any dimension. In the finance literature, these processes are known as the Cox–Ingersoll–Ross processes.

Another remarkable connection between linear combinations of reflecting BM and its local time with beta variables (recall Sect. 1.7) is:

Theorem 11.1.6 (F. Petit). *The following identity holds:*

$$\int_0^1 ds\ \mathbf{1}_{(|B_s|-\mu L_s \leq 0)} \overset{(law)}{=} Z_{\frac{1}{2},\frac{1}{2}\mu}. \tag{11.1.3}$$

If $\mu = 1$, $Z_{\frac{1}{2},\frac{1}{2}}$ is an arcsine variable.

Remark 11.1.7. To recover the case $\mu = 1$, it suffices to recall that Lévy's identity (or Tanaka formula!) implies that $(|B_s| - L_s; s \geq 0)$ is a BM.

11.2   On the Joint Laws of (S_b, I_b, L_b) and (S_1, I_1, L_1)

(a) Let b be a Brownian bridge, L_b be its local time at 0 up to time 1 and

$$S_b = \sup_{s \le 1} b(s), \quad I_b = -\inf_{s \le 1} b(s)$$

Theorem 11.2.1. *Let N be a standard normal r.v. independent of b. Then*

$$P\big(|N|S_b \le x, |N|I_b \le y, |N|L_b \in dl\big) = \exp\Big(-\frac{l}{2}(\coth x + \coth y)\Big)dl.$$

Undoubtedly, this formula would become much more complicated without the $|N|$ factor.

Proof. Let T be an exponential time with parameter $1/2$ independent of b, then

$$S_{g_T} \overset{\text{(law)}}{=} \sqrt{g_T}\,S_b \overset{\text{(law)}}{=} |N|S_b.$$

Such relations hold jointly for I_{g_T} and $L_T = L_{g_T}$, i.e.

$$(S_{g_T}, I_{g_T}, L_T) \overset{\text{(law)}}{=} (|N|S_b, |N|I_b, |N|L_b).$$

On the other hand, from the identity (10.2.3) enriched with a local time term, we have:

$$P(S_{g_T} \le x, I_{g_T} \le y, L_T \in dl) = E\big[e^{-\tau_l/2}\mathbf{1}_{(S_{\tau_l} \le x, I_{\tau_l} \le y)}\big]dl. \qquad (11.2.1)$$

We remark that:

$$(S_{\tau_\lambda} \le x) = (S_{\tau_{\lambda^-}} \le x) \cap (S_{(\tau_{\lambda^-}, \tau_\lambda)} \le x);$$

from which it is easily deduced with the help of the additive formula (A) that the quantity (11.2.1) equals:

$$\exp\Big(-l\int \mathbf{n}(d\varepsilon)\big(1 - e^{-V/2}\mathbf{1}_{(S_V \le x)(I_V \le y)}\big)\Big)dl.$$

Thus, to prove the theorem, it is enough to show that

$$\int \mathbf{n}(d\varepsilon)\big(1 - e^{-V/2}\mathbf{1}_{(S_V \le x)(I_V \le y)}\big) = \frac{1}{2}(\coth x + \coth y). \qquad (11.2.2)$$

We decompose $\mathbf{n}$ into $\mathbf{n}_+ + \mathbf{n}_-$ and then we compute

$$\int \mathbf{n}_+(d\varepsilon)\left(1 - e^{-V/2}\mathbf{1}_{(S_V \leq x)}\right)$$

$$= \mathbf{n}_+(S_V \geq x) + \int_{S_V \leq x} \mathbf{n}_+(d\varepsilon)(1 - e^{-V/2})$$

$$= \frac{1}{2}\int_x^\infty \frac{dm}{m^2} + \frac{1}{2}\int_0^x \left(1 - \left(\frac{m}{\sinh m}\right)^2\right)\frac{dm}{m^2} = \frac{1}{2}\coth x.$$

By symmetry, the term involving $\mathbf{n}_-$ is equal to $\frac{1}{2}\coth y$.   □

Some Consequences:

(i) *Computation of Moments*: We have

$$E[S_b^n I_b^m] = \frac{1}{E[|N|^{n+m}]} E\left[(|N|S_b)^n(|N|I_b)^m\right]. \qquad (11.2.3)$$

From Theorem 11.2.1, we deduce:

$$P\left(|N|S_b \leq x, |N|I_b \leq y\right) = \int dl \exp\left(-\frac{l}{2}(\coth x + \coth y)\right)$$

$$= \frac{2}{\coth x + \coth y} = \frac{(2\sinh x)(\sinh y)}{\sinh(x+y)}; \qquad (11.2.4)$$

hence, the formula

$$P\left(|N|S_b \in dx, |N|I_b \in dy\right) = \frac{4\sinh x \sinh y}{(\sinh(x+y))^3}dxdy$$

which yields a double integral formula for (11.2.3).
    As another consequence, we note that

$$4\int_0^\infty \int_0^\infty dxdy\,\frac{\sinh x \sinh y}{(\sinh(x+y))^3} = 1.$$

Letting $x + y = t$ gives:

$$4\int_0^\infty \frac{dt}{(\sinh t)^3}\int_0^t dx(\sinh x)(\sinh(t - x)) = 1,$$

thus showing clearly the convergence near $t = 0$.

(ii) *Law of* $M_b = \sup_{u \leq 1} |b(u)|$: We have, from Theorem 11.2.1, or rather from (11.2.4):

$$P(|N|M_b \leq x) = P(|N|S_b \leq x, |N|I_b \leq x) = \tanh x.$$

Consequently, we obtain the distribution:

$$P(|N|M_b \in dx) = \frac{dx}{(\cosh x)^2} \quad \text{on } \mathbb{R}_+;$$

$$P(NM_b \in dy) = \frac{dy}{(\cosh y)^2} \quad \text{on } \mathbb{R}$$

whose Fourier transform

$$\frac{\pi\lambda/2}{\sinh(\pi\lambda/2)}$$

coincides with the Laplace transform of $T^{(3)}_{\pi/2}$ taken in $\frac{\lambda^2}{2}$ where

$$T^{(3)}_{\pi/2} = \inf \left\{ t \geq 0 : R^{(3)}_t = \frac{\pi}{2} \right\}.$$

We conclude that $M_b^2 \overset{\text{(law)}}{=} T^{(3)}_{\pi/2}$.

(iii) Using the same kind of argument as in (ii) above, we get:

$$E \left[\exp \left(-\frac{\lambda^2}{2}(S_b + I_b)^2 \right) \right] = \left(\frac{(\pi\lambda)/2}{\sinh(\pi\lambda)/2} \right)^2.$$

On the other hand, from Vervaat's representation, discussed in [5], we obtain:

$$S_b + I_b \overset{\text{(law)}}{=} \sup_{s \leq 1} r(s)$$

where $(r(s); s \leq 1)$ denotes a BES(3) bridge.

We end up with Chung's identity

$$\left(\sup_{s \leq 1} r(s) \right)^2 \overset{\text{(law)}}{=} T^{(3)}_{\pi/2} + \tilde{T}^{(3)}_{\pi/2} \overset{\text{(law)}}{=} M_b^2 + \tilde{M}_b^2. \qquad (11.2.5)$$

Remark 11.2.2. This identity is related to the functional equation for Riemann's ζ function (see Sect. 11.6).

(b) With the help of the previous discussion of the law of (S_b, I_b, L_b), we are now able to obtain some analogous results for Brownian motion B itself.

We first prove the following formula

$$P\left(\sqrt{2T}\, S_1 \le x,\, \sqrt{2T}\, I_1 \le y\right) = 1 - \frac{\sinh x + \sinh y}{\sinh(x + y)}, \qquad (11.2.6)$$

where T denotes an exponential variable, with mean 1, independent of B. We check that formula (11.2.6) above agrees with formula (1.1.15.2) in Part II of Borodin–Salminen [4]. We also note that replacing T by t yields a theta-like expression (formula (1.1.15.4) in [4]).

Proof of (11.2.6). From the scaling property of the Brownian motion, we deduce that the LHS of (11.2.6) is equal to

$$P\left(S_{\tilde{T}} \le x, I_{\tilde{T}} \le y\right) = P(\tilde{T} \le T_{-y} \wedge T_x) = 1 - E\left[\exp\left(-\frac{1}{2}(T_{-y} \wedge T_x)\right)\right]$$

where $\tilde{T}$ is an independent exponential time, with mean 2, and $T_a = \inf\{t : B_t = a\}$.   □

Formula (11.2.6) now follows from the well-known formulae

$$E\left[\exp\left(-\frac{\lambda^2}{2} T_x\right); T_x < T_{-y}\right] = \frac{\sinh(\lambda y)}{\sinh(\lambda(x + y))}$$

$$E\left[\exp\left(-\frac{\lambda^2}{2} T_{-y}\right); T_{-y} < T_x\right] = \frac{\sinh(\lambda x)}{\sinh(\lambda(x + y))}$$

We now show how to obtain the trivariate distribution of (S_1, I_1, L_1), via that of $(\sqrt{2T}\, S_1, \sqrt{2T}\, I_1, \sqrt{2T}\, L_1)$. We first remark that, thanks to the independence of the pre-$g_{\tilde{T}}$ and post-$g_{\tilde{T}}$ processes, we have:

$$P\left(\sqrt{2T}\, S_1 \le x,\, \sqrt{2T}\, I_1 \le y,\, \sqrt{2T}\, L_1 \in dl\right) \qquad (11.2.7)$$
$$= P\left(|N|S_b \le x, |N|I_b \le y, |N|L_b \in dl\right) P\left(S_{(g_{\tilde{T}},\tilde{T})} \le x, I_{(g_{\tilde{T}},\tilde{T})} \le y\right).$$

We then remark that:

$$P\left(S_{(g_{\tilde{T}},\tilde{T})} \le x, I_{(g_{\tilde{T}},\tilde{T})} \le y\right) = \phi(x) + \phi(y),$$

where

$$\phi(x) = P\left(S_{(g_{\tilde{T}},\tilde{T})} \le x, B_{\tilde{T}} > 0\right) = \frac{1}{2} P\left(\sup_{g_{\tilde{T}} \le t \le \tilde{T}} |B_t| \le x\right).$$

Using formulae (11.2.6) and (11.2.7) in conjunction, as well as:

$$P(\sup_{t \leq \tilde{T}}|B_t| \leq x) = P(\sup_{t \leq g_{\tilde{T}}}|B_t| \leq x)2\phi(x)$$

we first obtain:

$$\phi(x) = \frac{1}{2}\tanh\left(\frac{x}{2}\right),$$

which, finally gives, thanks to (11.2.7) again:

$$P\left(\sqrt{2T}\,S_1 \leq x, \sqrt{2T}\,I_1 \leq y, \sqrt{2T}\,L_1 \in dl\right) \tag{11.2.8}$$
$$= \frac{1}{2}\left(\tanh\frac{x}{2} + \tanh\frac{y}{2}\right)\exp\left(-\frac{l}{2}(\coth x + \coth y)\right)dl.$$

Exercise 11.2.3. With the help of the above expression of ϕ, and the knowledge of the law of M_b obtained in (ii), compare the distributions of M_b and $\sup_{s \leq 1} m_s$, where m denotes the meander. (Answer: $M_b \overset{(\text{law})}{=} 2\sup_{s \leq 1} m(s)$).

Exercise 11.2.4. Show that formulae (11.2.6) and (11.2.8) agree, after integration of (11.2.8) in (dl).

11.3   Knight's Identity in Law

Theorem 11.3.1. *(Knight). Let $M_t = \sup_{s \leq t}|B_s|$. The following identity in law holds*

$$K \equiv \frac{\tau_1}{M_{\tau_1}^2} \overset{(\text{law})}{=} T_2^{(3)}.$$

Proof. We shall show that

$$E\left[\exp\left(-\frac{\lambda^2}{2}K\right)\right] = \frac{2\lambda}{\sinh 2\lambda} = \left(\frac{\lambda}{\sinh \lambda}\right)^2 \frac{1}{\lambda \coth \lambda}.$$

We give two different proofs.

First proof.
   As we have seen previously using excursion theory (see (11.2.2)):

$$E\left[\exp\left(-\frac{\tau_l}{2}\right)\mathbf{1}_{(M_{\tau_l} \leq x)}\right] = \exp(-l\coth x).$$

By scaling, we get:

$$E\left[\exp\left(-\frac{\tau_l}{2}\right)\mathbf{1}_{(M_{\tau_l}\le x)}\right] = E\left[\exp\left(-\frac{l^2\tau_1}{2}\right)\mathbf{1}_{(M_{\tau_1}\le\frac{x}{l})}\right].$$

Replacing x by (xl), we obtain:

$$E\left[\exp\left(-\frac{l^2\tau_1}{2}\right)\mathbf{1}_{(M_{\tau_1}\le x)}\right] = \exp(-l\coth(xl));$$

thus,

$$E\left[\exp\left(-\frac{l^2\tau_1}{2}\right)\mathbf{1}_{(M_{\tau_1}\in dx)}\right] = \exp(-l\coth(xl))\frac{l^2 dx}{(\sinh(xl))^2}.$$

As a consequence, we obtain

$$E\left[\exp\left(-\frac{l^2\tau_1}{2(M_{\tau_1})^2}\right)\right] = \int_0^\infty dx\exp\left(-\frac{l}{x}\coth l\right)\left(\frac{l}{x}\right)^2\frac{1}{(\sinh l)^2}$$

$$= \left(\frac{l}{\sinh l}\right)^2\left(\frac{1}{l\coth l}\right) = \frac{2l}{\sinh(2l)}.$$

$\square$

Second proof.
    We use the Ray–Knight theorem for $(L^x_{\tau_1}; x \ge 0)$. Let $B^+_t = \beta_{A^+_t}$ where $(\beta_u; u \ge 0)$ is a reflected BM. Then,

$$K \equiv \frac{\tau_1}{(\sup_{t\le\tau_1}|B_t|)^2} \overset{(\text{law})}{=} \frac{\tau_1^+}{(\sup_{t\le\tau_1}B_t)^2} \equiv \frac{\int_0^\infty dx L^x_{\tau_1}}{S^2_{\tau_1}}.$$

We note that $S_{\tau_1} = \inf\{x : L^x_{\tau_1} = 0\}$. By the Ray–Knight theorem (Theorem 2.2.4), $(L^x_{\tau_1}; 0 \le x \le S_{\tau_1})$ is a BESQ(0) considered up to its first hitting time of 0. We use a generalization of Williams' time reversal argument (See Revuz–Yor [8]) to write:

$$K \overset{(\text{law})}{=} \frac{1}{(\gamma_1^{(4)})^2}\int_0^{\gamma_1^{(4)}} dy\,(R_y^{(4)})^2 = \int_0^1 d\xi\left(\frac{1}{\sqrt{\gamma_1^{(4)}}}R^{(4)}_{\xi\gamma_1^{(4)}}\right)^2$$

where $(R_y^{(4)}; y \le \gamma_1^{(4)})$ denotes a BES(4) process up to $\gamma_1^{(4)}$, its last hitting time of 1.

We now use the following formula, due to Lévy, for $d = 4$,

$$E_0^{(d)}\left[\exp\left(-\frac{\lambda^2}{2}\int_0^1 d\xi R_\xi^2\right)\Big| R_1 = a\right] = \left(\frac{\lambda}{\sinh \lambda}\right)^{d/2} \exp\left(-\frac{a^2}{2}(\lambda \coth \lambda - 1)\right)$$

$$(11.3.1)$$

whence, with the help of formula (8.1.1), we obtain, for $d = n = 4$:

$$E\left[\exp\left(-\frac{\lambda^2}{2}K\right)\right]$$

$$= \left(\frac{\lambda}{\sinh \lambda}\right)^2 \int_0^\infty a\, da \exp\left(-\frac{a^2}{2}(\lambda \coth \lambda - 1)\exp\left(-\frac{a^2}{2}\right)\right)$$

$$= \left(\frac{\lambda}{\sinh \lambda}\right)^2 \left(\frac{1}{\lambda \coth \lambda}\right) = \frac{2\lambda}{\sinh(2\lambda)}.$$

$\square$

11.4   The Földes–Révész Identity

Theorem 11.4.1. *Let $0 < a \leq 1$ The following identity in law holds*

$$\int_0^\infty dx \mathbf{1}_{(0 < L_{\tau_1}^x < a)} \stackrel{(law)}{=} T_{\sqrt{a}}^{(2)}$$

where $T_y^{(2)} \equiv \inf\{t \geq 0 : R_t^{(2)} = y\}$.

Proof. From the first Ray–Knight theorem (Theorem 2.2.4), we have

$$\int_0^\infty dx \mathbf{1}_{(0 < L_{\tau_1}^x < a)} = \int_0^{T_0} dx \mathbf{1}_{(0 < X_x < a)}$$

where $(X_x = L_{\tau_1}^x; x \geq 0)$ is a $\text{BESQ}_1(0)$ process. Using the same time reversal argument as in Sect. 11.3, we obtain:

$$\int_0^{T_0} dx \mathbf{1}_{(0 < X_x < a)} \stackrel{(law)}{=} \int_0^{\hat{\gamma}_1^{(4)}} dy \mathbf{1}_{(0 < \hat{X}_y < a)}$$

where $(\hat{X}_y; y > 0)$ is a $\text{BESQ}_0(4)$ process; because $a \leq 1$, the last expression above is equal to:

$$\int_0^\infty dy \mathbf{1}_{(\hat{X}_y < a)} = \int_0^\infty dy \mathbf{1}_{(R_y^{(4)} < \sqrt{a})}$$

and $(R_y^{(4)}; y \geq 0)$ is a $\text{BES}_0(4)$ process.

A theorem of Ciesielski–Taylor [10] and Yor [12] asserts that

$$\int_0^\infty dy\,1_{\left(R_y^{(d+2)}<x\right)} \overset{(\text{law})}{=} T_x^{(d)}$$

which implies the result for $d = 2$.   □

The following proposition employs similar arguments as above.

Proposition 11.4.2. *Let $a > 0$, and $(\tilde{B}_u; u \le \tilde{T}_0)$ be a BM starting from 1, and considered up to $\tilde{T}_0 = \inf\{t : \tilde{B}_t = 0\}$. The following four pairs of random variables are identically distributed*

(a) $\left(\int_0^\infty dx\,1_{(0<L_{\tau_1}^x<a)}, \tau_1^+ \equiv A_{\tau_1}^+ \right)$;

(b) $\left(\int_0^{\tilde{T}_0} \frac{ds}{4\tilde{B}_s}1_{(\tilde{B}_s<a)}, \frac{1}{4}\tilde{T}_0 \right)$;

(c) $\left(\int_0^{\gamma_1^{(3)}} \frac{ds}{4R_s^{(3)}}1_{(R_s^{(3)}<a)}, \frac{1}{4}\gamma_1^{(3)} \right)$;

(d) $\left(\int_0^{\gamma_1^{(4)}} dy\,1_{(R_y^{(4)}<\sqrt{a})}, \int_0^{\gamma_1^{(4)}} dy\left(R_y^{(4)}\right)^2 \right)$.

Corollary 11.4.3. *As $a \to \infty$, the above four pairs of random variables converge respectively to the four identically distributed pairs, i.e. (S_{τ_1}, τ_1^+) is distributed as:*

$$\left(\int_0^{\tilde{T}_0} \frac{ds}{4\tilde{B}_s}, \frac{1}{4}\tilde{T}_0 \right) \overset{(\text{law})}{=} \left(\int_0^{\gamma_1^{(3)}} \frac{ds}{4R_s^{(3)}}, \frac{1}{4}\gamma_1^{(3)} \right) \overset{(\text{law})}{=} \left(\gamma_1^{(4)}, \int_0^{\gamma_1^{(4)}} dy\left(R_y^{(4)}\right)^2 \right).$$

Remark 11.4.4. Note that: $S_{\tau_1} = \int_0^\infty dx\,1_{(0<L_{\tau_1}^x<\infty)}$.

Corollary 11.4.3 gives some explanations for the appearance of Lévy's formulae in the computations of Laplace transforms of distributions of functionals which are, a priori, very different from the stochastic area of planar BM; see, e.g. Sect. 11.5.

Proof of Proposition 11.4.2.

(a) $\overset{(\text{law})}{=}$ (d) is a consequence of the time reversal from $\text{BESQ}_1(0)$ to $\text{BESQ}_0(4)$.

(a) $\overset{(\text{law})}{=}$ (b) follows from the time change $X_x = \tilde{B}_{4\int_0^x dy X_y}$,
where $(X_x \equiv L_{\tau_1}^x ; x \ge 0)$ denotes a $\text{BESQ}_1(0)$ process.

(b) $\overset{(\text{law})}{=}$ (c) is a consequence of Williams' time reversal theorem.   □

Exercise 11.4.5. Prove (c) $\overset{(\text{law})}{=}$ (d) using the transformation BES(3) → BES(4) by time substitution.

11.5   Cauchy Principal Value of Brownian Local Times

Let $H_t = \int_0^t \frac{ds}{B_s}$ and T be an exponential random time with parameter $1/2$, independent of $(B_s ; s \geq 0)$. We write

$$H_T = H_T^- + H_T^+,$$

where $H_T^- = H_{g_T}$ and $H_T^+ = H_T - H_{g_T}$.

Proposition 11.5.1. *There is the identity:*

$$E\left[\exp\left(\frac{i\lambda}{\pi} H_{\tau_l} - \frac{\theta^2}{2} \tau_l \right) \right] = \exp\left(-\lambda l \coth\left(\frac{\lambda}{\theta} \right) \right).$$

We check that when $\theta \to 0$, the expression on the RHS converges to $\exp(-l|\lambda|)$ and thus $(\frac{1}{\pi} H_{\tau_l} ; l \geq 0)$ is standard Cauchy process; when $\lambda \to 0$, the same expression converges to $\exp(-l\theta)$ and therefore, we recover that $(\tau_l ; l \geq 0)$ is a stable subordinator of index $(1/2)$.

Theorem 11.5.2.

(1) The variables H_T^- and H_T^+ are independent.
(2) Moreover, we have:

$$E\left[\exp\left(\frac{i\lambda}{\pi} H_T^- \right) \right] = \frac{\tanh \lambda}{\lambda}; \quad E\left[\exp\left(\frac{i\lambda}{\pi} H_T^+ \right) \right] = \frac{\lambda}{\sinh \lambda}$$

$$E\left[\exp\left(\frac{i\lambda}{\pi} H_T \right) \right] = \frac{1}{\cosh \lambda}$$

and

$$E\left[\exp\left(\frac{i\lambda}{\pi} H_T \right) \Big| L_T = l \right] = \frac{\lambda}{\sinh \lambda} \exp\left(-l(\lambda \coth \lambda - 1) \right).$$

Consequences*: Thanks to Lévy's area formula (11.3.1) for $d = 2$, the following identity in law holds*

$$\left(\frac{1}{\pi} H_T, L_T \right) \overset{(law)}{=} \left(\int_0^1 B_s d\hat{B}_s - \hat{B}_s dB_s, \frac{1}{2} R_1^2 \right),$$

where $R_1^2 = B_1^2 + \hat{B}_1^2$ and B and $\hat{B}$ are two independent BM's starting from 0.

Remark 11.5.3. One can obtain the law of H_T^+ from the law of $(\tilde{H}_{\tilde{T}_0}, \tilde{T}_0)$ where $\tilde{H}_{\tilde{T}_0} = \int_0^{\tilde{T}_0} \frac{ds}{\tilde{B}_s}$.

Note that

$$H_T^+ \overset{\text{(law)}}{=} \sqrt{T - g_T} \, H_m \, \text{sgn}(B_1) \overset{\text{(law)}}{=} N H_m$$

where $H_m = \int_0^1 \frac{ds}{m(s)}$ and N is a standard normal r.v. independent of H_m.
Analogously,

$$H_T^- \overset{\text{(law)}}{=} \sqrt{g_T} \, H_b \overset{\text{(law)}}{=} N H_b$$

where N is a standard normal r.v. independent of $H_b = \int_0^1 \frac{ds}{b(s)}$.

Corollary 11.5.4.

$$E\left[\exp\left(-\frac{\lambda^2}{2}\left(\frac{1}{\pi}H_b\right)^2\right)\right] = \frac{\tanh \lambda}{\lambda}$$

$$E\left[\exp\left(-\frac{\lambda^2}{2}\left(\frac{1}{\pi}H_m\right)^2\right)\right] = \frac{\lambda}{\sinh \lambda} \quad \left(= E\left[\exp\left(-\frac{\lambda^2}{2}T_1^{(3)}\right)\right]\right)$$

whence

$$H_m \overset{\text{(law)}}{=} 2 \sup_{s \le 1} m(s) \overset{\text{(law)}}{=} \sup_{s \le 1} |b(s)|, \quad \text{(see Exercise 11.2.3)}$$

Moreover,

$$E\left[\exp\left(-\frac{\lambda^2}{2}\left(\frac{1}{\pi}H_r\right)^2\right)\right] = \left(\frac{\lambda}{\sinh \lambda}\right)^2$$

whence

$$H_r = \int_0^1 \frac{ds}{r(s)} \overset{\text{(law)}}{=} 2 \sup_{s \le 1} r(s),$$

thanks to Chung's identity (11.2.5).

Remark 11.5.5. This last identity in law is closely related to the functional equation for the Riemann ζ function which is discussed in the next section.

11.6   The Agreement Formula and the Functional Equation of the Riemann ζ Function

To begin with, recall that the Riemann ζ function may be defined by:

$$\zeta(s) = \sum_{n=1}^{\infty} \frac{1}{n^s} \quad \text{for } s \in \mathbb{C} \quad \text{such that } \text{Re}(s) > 1.$$

Then, ζ extends analytically to $\mathbb{C}$ as a meromorphic function with a simple pole at $s = 1$. Moreover, ζ satisfies the functional equation

$$\xi(s) = \xi(1 - s) \tag{11.6.1}$$

where

$$\xi(s) = \frac{s(s-1)}{2} \Gamma\left(\frac{s}{2}\right) \pi^{-s/2} \zeta(s).$$

We will describe ξ as a moment function. Let

$$Z = \frac{\pi}{2}\hat{T} \quad \text{where} \quad \hat{T} = T_1^{(3)} + \tilde{T}_1^{(3)},$$

and $T_1^{(3)}$ and $\tilde{T}_1^{(3)}$ denote the first hitting times of 1 for two independent BES(3) processes. Then

$$2\xi(2s) = E[Z^s]$$

and (11.6.1) can be written in terms of Z as

$$E[Z^s] = E[Z^{\frac{1}{2}-s}]$$

or equivalently, for every $f : \mathbb{R}_+ \to \mathbb{R}_+$

$$E[f(Z)] = E\left[f\left(\frac{1}{Z}\right)\sqrt{Z}\right],$$

and from the definition of Z, we get the following equation, which is equivalent to (11.6.1)

$$E\left[f\left(\frac{\pi^2}{4}\hat{T}\right)\right] = \sqrt{\frac{\pi}{2}} E\left[f\left(\frac{1}{\hat{T}}\right)\sqrt{\hat{T}}\right]. \tag{11.6.2}$$

Recall the result shown in Sect. 11.5, or rather in (11.2.5):

$$\left(\sup_{s \le 1} r(s)\right)^2 \stackrel{(\text{law})}{=} \frac{\pi^2}{4}\hat{T} = T_{\pi/2}^{(3)} + \tilde{T}_{\pi/2}^{(3)},$$

hence (11.6.2), and consequently (11.6.1), are both equivalent to

$$E\left[f\left(\left(\sup_{s \le 1} r(s)\right)^2\right)\right] = \sqrt{\frac{\pi}{2}} E\left[f\left(\frac{1}{\hat{T}}\right)\sqrt{\hat{T}}\right]. \tag{11.6.3}$$

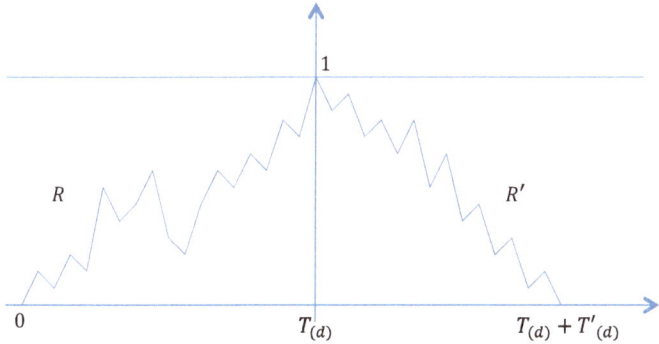

Fig. 11.1 Graph for Theorem 11.6.1

Now this identity will appear as an application of the following general agreement formula between the Bessel bridge and two Bessel processes put back to back (Fig. 11.1).

Theorem 11.6.1 (The agreement formula; Pitman–Yor [19]). *Let $d > 0$ and denote by R and R' two $BES_0(d)$ processes. Let $T_{(d)}$ and $T'_{(d)}$ be their respective first hitting times of 1. We define:*

$$\rho_u = \begin{cases} R_u, & \text{if } u \leq T_{(d)} \\ R'_{(T_{(d)}+T'_{(d)}-u)}, & \text{if } T_{(d)} \leq u \leq T_{(d)} + T'_{(d)} \end{cases}$$

Denote by $\tilde{\rho}$ the process $\rho^{[0,T_{(d)}+T'_{(d)}]}$ defined on $[0,1]$. If $(r_v; v \leq 1)$ is a d-dimensional Bessel bridge, the following absolute continuity relation holds

$$E[F(r_v; v \leq 1)] = 2^{\frac{d}{2}-1} \Gamma\left(\frac{d}{2}\right) E\left[F(\tilde{\rho}_v; v \leq 1)(T_{(d)} + T'_{(d)})^{\frac{d}{2}-1}\right].$$

Remark 11.6.2. If $d = 2$, the agreement formula simply becomes:

$$\left(r(v); v \leq 1\right) \overset{\text{(law)}}{=} \left(\tilde{\rho}_v; v \leq 1\right).$$

Corollary 11.6.3. *There is a unique time $\theta_{(d)}$ at which r reaches its maximum. For $f : \mathbb{R}_+^2 \to \mathbb{R}_+$ one has*

$$E^{(d)}\left[f\left(\left(\sup_{s\leq 1} r(s)\right)^2, \theta_{(d)}\right)\right]$$

$$= 2^{\frac{d}{2}-1} \Gamma\left(\frac{d}{2}\right) E\left[f\left(\frac{1}{T_{(d)}+T'_{(d)}}, \frac{T_{(d)}}{T_{(d)}+T'_{(d)}}\right)(T_{(d)} + T'_{(d)})^{\frac{d}{2}-1}\right]$$

which yields (11.6.3) when $d = 3$, and f depends only on the first variable.

Proof of the agreement formula in the case $d = 3$. We use the Itô–Williams representations of $\mathbf{n}$ to calculate

$$I \equiv \int \mathbf{n}(d\varepsilon) E\left[\varphi(V) F\left(\frac{1}{\sqrt{V}} \varepsilon_{uV}; u \le 1\right)\right].$$

□

(a) Conditioning with respect to V,

$$I = \int_0^\infty \frac{dv}{\sqrt{2\pi v^3}} \varphi(v) E[F(r_u; u \le 1)].$$

(b) Conditioning with respect to M, and using the scaling property,

$$I = \int_0^\infty \frac{dm}{m^2} E\left[\varphi\left(m^2(T_1 + T_1')\right) F(\tilde{\rho}_u; u \le 1)\right].$$

Using Fubini's theorem and the usual change of variables, one gets the desired formula.

11.7 On Ranked Lengths of Excursions

For $(t > 0)$ (possibly random) we consider the lengths of excursions away from 0 for BM,

$$V_1(t) > V_2(t) > \cdots > V_n(t) > \cdots$$

including $t - g_t$. To get a better understanding of the invariance results in the derivation of the second arcsine law (Theorem 4.2.1), the following theorem was obtained.

Theorem 11.7.1 (Pitman–Yor [24]). *The law of*

$$\frac{\mathbf{V}(T)}{T} \equiv \left(\frac{V_1(T)}{T}, \frac{V_2(T)}{T}, \cdots\right)$$

is the same for

$$(i)\ T = t, \quad \text{and for}\ (ii)\ T = \tau_s$$

It is interesting, although difficult, to obtain the finite-dimensional marginals laws of this infinite-dimensional variable. Note first that, by scaling property, the one-dimensional marginals can be understood with the help of the following sequence of stopping times

$$H_a^{(1)} = \inf\{t : V_1(t) \geq a\}; H_a^{(2)} = \inf\{t : V_2(t) \geq a\}; \ldots;$$
$$H_a^{(n)} = \inf\{t : V_n(t) \geq a\}$$

For each $n \geq 1$, the scaling property yields:

$$V_n(1) \overset{(\text{law})}{=} \frac{1}{H_1^{(n)}} \tag{11.7.1}$$

and we can easily obtain the Laplace transform of $H_1^{(n)}$, for any fixed n, thanks to the following argument:

(i) Let $n = 1$, from the independence between $\mathcal{F}_{g_t}$ and the meander

$$\left(m_t(u) \equiv \frac{1}{\sqrt{t - g_t}} \left| B_{g_t + u(t - g_t)} \right| ; u \leq 1 \right),$$

we obtain an explicit expression for the projection on $\mathcal{F}_{g_t}$ of the $\mathcal{F}_t$-martingale

$$M_t \equiv \cosh(\lambda |B_t|) \exp\left(-\frac{\lambda^2 t}{2}\right).$$

Indeed,

$$E[M_t | \mathcal{F}_{g_t}] = \psi(\lambda \sqrt{t - g_t}) \exp\left(-\frac{\lambda^2 t}{2}\right)$$

where

$$\psi(\lambda) = E[\cosh(\lambda m_1)] = \frac{1}{2}(\varphi_+(\lambda) + \varphi_-(\lambda))$$

and

$$\varphi_+(\lambda) = E[\exp(\lambda m_1)], \quad \varphi_-(\lambda) = E[\exp(-\lambda m_1)].$$

More explicitly,

$$\varphi_+(\lambda) = \sqrt{2\pi}\lambda \exp\left(\frac{\lambda^2}{2}\right) + \varphi_-(\lambda)$$

$$\varphi_-(\lambda) = \frac{1}{2}\int_0^\infty \frac{dy}{(1 + y)^{3/2}} \exp\left(-\frac{\lambda^2 y}{2}\right)$$

and therefore,

$$\psi(\lambda) = \sqrt{\frac{\pi}{2}}\lambda \exp\left(\frac{\lambda^2}{2}\right) + \frac{1}{2}\int_0^\infty \frac{dy}{(1 + y)^{3/2}} \exp\left(-\frac{\lambda^2 y}{2}\right).$$

Using the optional sampling theorem, we get a formula for the Laplace transform of $H_a^{(1)}$:

$$E\left[\exp\left(-\frac{\lambda^2}{2}H_a^{(1)}\right)\right] = \frac{1}{\psi(\lambda\sqrt{a})}.$$

(ii) Let now $n > 1$. It is easy to see that for a fixed a, $(H_a^{(n)}; n \geq 1)$ is the sequence of stopping times obtained, starting with $H_a^{(1)}$, via the following iteration procedure:

$$H_a^{(n)} = \inf\{u \geq d_{H_a^{(n-1)}}; u - g_u = a\}$$

Then we have the following (Fig. 11.2):

Proposition 11.7.2.

(1) The random variables $H^{(1)}, H^{(2)} - H^{(1)}, \ldots, H^{(n)} - H^{(n-1)}$ are independent.
(2) For $n \geq 1$, the variables $H^{(n)} - H^{(n-1)}$ are identically distributed.
(3) The variables $\left(B_{H^{(1)}}, H^{(1)}, d_{H^{(1)}} - H^{(1)}, H^{(2)} - d_{H^{(1)}}\right)$ are independent and
$B_{H^{(1)}} \overset{(law)}{=} \eta m_1$ where η is a Bernoulli variable independent of m_1.

Proof. We only need to show (3), which follows from the independence of $B_{H^{(1)}}$ and $H^{(1)}$. This is a consequence of the fact that for each stopping time T with respect to $(\mathcal{F}_{g_t})$, the meander $(m_T(u); u \leq 1)$ is independent of $\mathcal{F}_{g_T}$. In this case, we have

$$T = H^{(1)} \quad \text{and} \quad T - g_T = 1$$

and thus $(|B_{g_{H^{(1)}}+u}|; u \leq 1)$ is a meander independent of $\mathcal{F}_{g_{H^{(1)}}}$, hence of $g_{H^{(1)}} \equiv H^{(1)} - 1$, hence of $H^{(1)}$.

We obtain thus the Laplace transform of $H^{(n+1)}$

$$E\left[\exp\left(-\frac{\lambda^2}{2}H^{(n+1)}\right)\right] = E\left[e^{-\frac{\lambda^2}{2}H^{(1)}}\right]\left(E\left[e^{-\frac{\lambda^2}{2}(H^{(2)}-H^{(1)})}\right]\right)^n.$$

$\square$

Remark 11.7.3. The one-dimensional identities in law in identity (11.7.1) cannot be extended to the k-dimensional marginals.

Following the development in the case $n = 1$, or more generally, our study in Sect. 7.6, we can consider several problems related to the process $(\sqrt{t - g_t}; t \geq 0)$ by analogy with the reflected BM, or related to the process $(\text{sgn}(B_t)\sqrt{t - g_t}; t \geq 0)$ by analogy with BM.

As an example, we consider the joint law of $V_1^+(b)$ and $V_1^-(b)$, and denote:

$$V_1(b) = \max\left(V_1^+(b), V_1^-(b)\right)$$

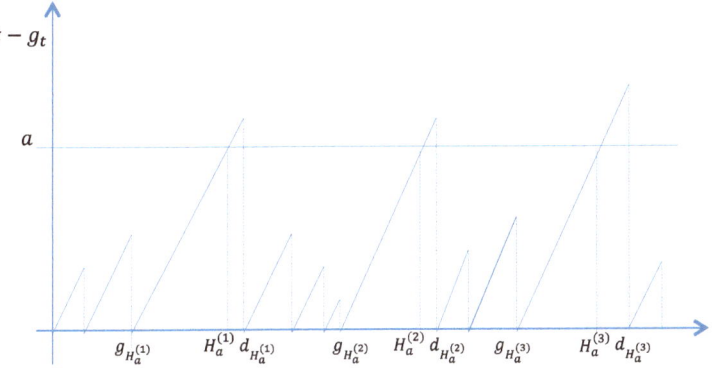

Fig. 11.2 Illustration for Proposition 11.7.2

Let T be an exponential random time with parameter $(1/2)$ independent of b, then

$$P\Big(|N|(V_1^+(b))^{1/2} \leq x, |N|(V_1^-(b))^{1/2} \leq y, |N|L_b \in dl\Big)$$

$$= P\Big((V_1^+(g_T))^{1/2} \leq x, (V_1^-(g_T))^{1/2} \leq y, L_T \in dl\Big)$$

$$= E\Big[\exp\Big(-\frac{\tau_l}{2}\Big)\mathbf{1}_{((V_1^+(\tau_l))^{1/2}\leq x,(V_1^-(\tau_l))^{1/2}\leq y)}\Big]dl$$

$$= \exp(-l\ p(x,y))dl$$

where

$$p(x,y) = \frac{1}{2}\Big(\int_0^\infty \frac{dv}{\sqrt{2\pi v^3}}\big(1 - e^{-v/2}\mathbf{1}_{(v \leq x^2)}\big) + \int_0^\infty \frac{dv}{\sqrt{2\pi v^3}}\big(1 - e^{-v/2}\mathbf{1}_{(v \leq y^2)}\big)\Big).$$

We remark that:

$$p(x,y) = \frac{1}{2}\big(c(x) + c(y)\big),$$

with

$$c(x) = 1 + \int_{x^2}^\infty \frac{dv}{\sqrt{2\pi v^3}}\big(e^{-v/2}\big).$$

The function c enjoys the following relations with φ_- and ψ, as defined above,

$$c(x) = 1 + \sqrt{\frac{2}{\pi}}\Big(\frac{e^{-x^2/2}}{x}\Big)\varphi_-(x) = \sqrt{\frac{\pi}{2}}\Big(\frac{e^{-x^2/2}}{x}\Big)\psi(x).$$

In particular, we obtain:

$$P\left(|N|(V_1(b))^{1/2} \in dx\right) = -\frac{c'(x)}{(c(x))^2}dx.$$

As a consequence, the following formula for the Laplace transform of $V_1(b)$:

$$E\left[\exp\left(-\frac{\lambda^2}{2}V_1(b)\right)\right] = \int_{-\infty}^{\infty} dx\, e^{i\lambda x}\, \frac{e^{-x^2/2}}{(\psi(x))^2}\sqrt{\frac{2}{\pi^3}}$$

holds.

Notes and Comments:

1. Section 11.4 shows that, thanks to the Ray–Knight theorems (Theorem 2.2.4) and the importance of time changes which relate different diffusions, Lévy's stochastic area formula may occur in various computations which, a priori, involve extremely different quantities than the stochastic area of two-dimensional Brownian motion.
2. Section 11.5 originates from Biane–Yor [14].

    The results have been generalized by Fitzsimmons and Getoor [15] to symmetric stable processes; Bertoin [16] shows how to relate the Fourier aspects of the Hilbert transform with Feynman–Kac's formula.

References

1. M. Émery, E. Perkins, La filtration de $B + L$. Z. Wahrsch. Verw. Gebiete **59**(3), 383–390 (1982)
2. J.-F. Le Gall, M. Yor, Excursions browniennes et carrés de processus de Bessel. C. R. Acad. Sci. Paris Sér. I Math. **303**(3), 73–76 (1986)
3. J. Pitman, M. Yor, A decomposition of Bessel bridges. Z. Wahrsch. Verw. Gebiete **59**(4), 425–457 (1982)
4. A.N. Borodin, P. Salminen, Handbook of Brownian motion—facts and formulae. *Probability and Its Applications* (Birkhäuser, Boston, 1996)
5. W. Vervaat, A relation between Brownian bridge and Brownian excursion. Ann. Probab. **7**(1), 143–149 (1979)
6. Ph. Biane, Sur un calcul de F. Knight. Séminaire de Probabilités, XXII. *Lecture Notes in Math.*, vol. 1321. (Springer, Berlin, 1988), pp. 190–196
7. F.B. Knight, Inverse local times, positive sojourns, and maxima for Brownian motion. Colloque Paul Lévy sur les Processus Stochastiques (Palaiseau, 1987). Astérisque 157–158, 233–247 (1988)
8. D. Revuz, M. Yor, Continuous martingales and Brownian motion. *Grundlehren der Mathematischen Wissenschaften [Fundamental Principles of Mathematical Sciences]*, vol. 293, 3rd edn. (Springer, Berlin, 1999)

9. P. Vallois, Sur la loi conjointe du maximum et de l'inverse du temps local du mouvement brownien: application à un théorème de Knight. Stochast. Stochast. Rep. **35**(3), 175–186 (1991)

10. Z. Ciesielski, S.J. Taylor, First passage times and sojourn times for Brownian motion in space and the exact Hausdorff measure of the sample path. Trans. Am. Math. Soc. **103**, 434–450 (1962)

11. A. Földes, P. Révész, On hardly visited points of the Brownian motion. Probab. Theor Relat. Field **91**(1), 71–80 (1992)

12. M. Yor, Une explication du théorème de Ciesielski-Taylor. Ann. Inst. H. Poincaré Probab. Stat. **27**(2), 201–213 (1991)

13. M. Yor, On an identity in law obtained by A. Földes and P. Révész. Ann. Inst. H. Poincaré Probab. Stat. **29**(2), 321–324 (1993)

14. Ph. Biane, M. Yor, Valeurs principales associées aux temps locaux browniens. Bull. Sci. Math. (2) **111**(1), 23–101 (1987)

15. P.J. Fitzsimmons, R.K. Getoor, On the distribution of the Hilbert transform of the local time of a symmetric Lévy process. Ann. Probab. **20**(3), 1484–1497 (1992)

16. J. Bertoin, On the Hilbert transform of the local times of a Lévy process. Bull. Sci. Math. **119**(2), 147–156 (1995)

17. J. Bertoin, Lévy processes. *Cambridge Tracts in Mathematics*, vol. 121. (Cambridge University Press, Cambridge, 1996)

18. P. Biane, J. Pitman, M. Yor, Probability laws related to the Jacobi theta and Riemann zeta functions, and Brownian excursions. Bull. Am. Math. Soc. (N.S.) **38**(4), 435–465 (2001)

19. J. Pitman, M. Yor, Decomposition at the maximum for excursions and bridges of one-dimensional diffusions. *Itô's Stochastic Calculus and Probability Theory*. (Springer, Tokyo, 1996), pp. 293–310

20. D. Williams, Brownian motion and the Riemann zeta-function. *Disorder in Physical Systems, Oxford Sci. Publ.* (Oxford University Press, Oxford, 1990), pp. 361–372

21. R.K. Getoor, Excursions of a Markov process. Ann. Probab. **7**(2), 244–266 (1979)

22. F.B. Knight, On the duration of the longest excursion. Seminar on stochastic processes, 1985 (Gainesville, Fla., 1985). *Progr. Probab. Statist.*, vol. 12. (Birkhäuser, Boston, 1986), pp. 117–147

23. M. Perman, Order statistics for jumps of normalised subordinators. Stochast. Process. Appl. **46**(2), 267–281 (1993)

24. J. Pitman, M. Yor, Arcsine laws and interval partitions derived from a stable subordinator. Proc. Lond. Math. Soc. (3) **65**(2), 326–356 (1992)

General References

1. J. Azéma, M. Yor, (eds.) Temps locaux. Astérisque 52–53 (1978)
2. A.N. Borodin, P. Salminen, Handbook of Brownian motion—facts and formulae. *Probability and Its Applications*, (Birkhäuser, Boston, 1996)
3. K.L. Chung, R.J. Williams, Introduction to stochastic integration, *Probability and Its Applications*, 2nd edn. (Birkhäuser, Boston, 1990)
4. N. Ikeda, S. Watanabe, Stochastic differential equations and diffusion processes. *North-Holland Mathematical Library*, vol. 24, 2nd edn. (North-Holland Publishing, Amsterdam, 1989)
5. I. Karatzas, S.E. Shreve, Brownian motion and stochastic calculus. *Graduate Texts in Mathematics*, vol. 113, 2nd edn. (Springer, Berlin, 1991)
6. F.B. Knight, Essentials of Brownian motion and diffusion. Math. Surv. 18 (1981)
7. B. Mallein, M. Yor, Temps locaux de semi-martingales continues et excursions browniennes. In preparation (2013)
8. R. Mansuy, M. Yor, Aspects of Brownian motion. *Universitext*. (Springer, Berlin, 2008)
9. P. Mörters, Y. Peres, *Brownian Motion*. (Cambridge University Press, Cambridge, 2010)
10. L.C.G. Rogers, D. Williams, Diffusions, Markov processes, and martingales. *Cambridge Mathematical Library, Itô Calculus*, Vol. 1. Reprint of the second edition (1994) (Cambridge University Press, Cambridge, 2000)
11. L.C.G. Rogers, D. Williams, Diffusions, Markov processes, and martingales. *Cambridge Mathematical Library, Itô Calculus*, Vol. 2. Reprint of the second (1994) edition. (Cambridge University Press, Cambridge, 2000)
12. D. Revuz, M. Yor, Continuous martingales and Brownian motion. *Grundlehren der Mathematischen Wissenschaften [Fundamental Principles of Mathematical Sciences]*, vol. 293, 3rd edn. (Springer, Berlin, 1999)
13. M. Yor, Some aspects of Brownian motion. Part I. Some special functionals. *Lectures in Mathematics ETH Zürich* (Birkhäuser, Basel, 1992)
14. M. Yor, Some aspects of Brownian motion. Part II. Some recent martingale problems. *Lectures in Mathematics ETH Zürich*. (Birkhäuser, Basel, 1997)

Index

J.-Y. Yen and M. Yor, *Local Times and Excursion Theory for Brownian Motion*,
Lecture Notes in Mathematics 2088, DOI 10.1007/978-3-319-01270-4,
© Springer International Publishing Switzerland 2013

LECTURE NOTES IN MATHEMATICS

 Springer

Edited by J.-M. Morel, B. Teissier; P.K. Maini

Editorial Policy (for the publication of monographs)

1. Lecture Notes aim to report new developments in all areas of mathematics and their applications - quickly, informally and at a high level. Mathematical texts analysing new developments in modelling and numerical simulation are welcome.

   Monograph manuscripts should be reasonably self-contained and rounded off. Thus they may, and often will, present not only results of the author but also related work by other people. They may be based on specialised lecture courses. Furthermore, the manuscripts should provide sufficient motivation, examples and applications. This clearly distinguishes Lecture Notes from journal articles or technical reports which normally are very concise. Articles intended for a journal but too long to be accepted by most journals, usually do not have this "lecture notes" character. For similar reasons it is unusual for doctoral theses to be accepted for the Lecture Notes series, though habilitation theses may be appropriate.

2. Manuscripts should be submitted either online at www.editorialmanager.com/lnm to Springer's mathematics editorial in Heidelberg, or to one of the series editors. In general, manuscripts will be sent out to 2 external referees for evaluation. If a decision cannot yet be reached on the basis of the first 2 reports, further referees may be contacted: The author will be informed of this. A final decision to publish can be made only on the basis of the complete manuscript, however a refereeing process leading to a preliminary decision can be based on a pre-final or incomplete manuscript. The strict minimum amount of material that will be considered should include a detailed outline describing the planned contents of each chapter, a bibliography and several sample chapters.

   Authors should be aware that incomplete or insufficiently close to final manuscripts almost always result in longer refereeing times and nevertheless unclear referees' recommendations, making further refereeing of a final draft necessary.

   Authors should also be aware that parallel submission of their manuscript to another publisher while under consideration for LNM will in general lead to immediate rejection.

3. Manuscripts should in general be submitted in English. Final manuscripts should contain at least 100 pages of mathematical text and should always include

   - a table of contents;
   - an informative introduction, with adequate motivation and perhaps some historical remarks: it should be accessible to a reader not intimately familiar with the topic treated;
   - a subject index: as a rule this is genuinely helpful for the reader.

   For evaluation purposes, manuscripts may be submitted in print or electronic form (print form is still preferred by most referees), in the latter case preferably as pdf- or zipped psfiles. Lecture Notes volumes are, as a rule, printed digitally from the authors' files. To ensure best results, authors are asked to use the LaTeX2e style files available from Springer's web-server at:

   ftp://ftp.springer.de/pub/tex/latex/svmonot1/ (for monographs) and
   ftp://ftp.springer.de/pub/tex/latex/svmultt1/ (for summer schools/tutorials).

Additional technical instructions, if necessary, are available on request from lnm@springer.com.

4. Careful preparation of the manuscripts will help keep production time short besides ensuring satisfactory appearance of the finished book in print and online. After acceptance of the manuscript authors will be asked to prepare the final LaTeX source files and also the corresponding dvi-, pdf- or zipped ps-file. The LaTeX source files are essential for producing the full-text online version of the book (see http://www.springerlink.com/openurl.asp?genre=journal&issn=0075-8434 for the existing online volumes of LNM). The actual production of a Lecture Notes volume takes approximately 12 weeks.

5. Authors receive a total of 50 free copies of their volume, but no royalties. They are entitled to a discount of 33.3 % on the price of Springer books purchased for their personal use, if ordering directly from Springer.

6. Commitment to publish is made by letter of intent rather than by signing a formal contract. Springer-Verlag secures the copyright for each volume. Authors are free to reuse material contained in their LNM volumes in later publications: a brief written (or e-mail) request for formal permission is sufficient.

Addresses:
Professor J.-M. Morel, CMLA,
École Normale Supérieure de Cachan,
61 Avenue du Président Wilson, 94235 Cachan Cedex, France
E-mail: morel@cmla.ens-cachan.fr

Professor B. Teissier, Institut Mathématique de Jussieu,
UMR 7586 du CNRS, Équipe "Géométrie et Dynamique",
175 rue du Chevaleret
75013 Paris, France
E-mail: teissier@math.jussieu.fr

For the "Mathematical Biosciences Subseries" of LNM:

Professor P. K. Maini, Center for Mathematical Biology,
Mathematical Institute, 24-29 St Giles,
Oxford OX1 3LP, UK
E-mail: maini@maths.ox.ac.uk

Springer, Mathematics Editorial, Tiergartenstr. 17,
69121 Heidelberg, Germany,
Tel.: +49 (6221) 4876-8259

Fax: +49 (6221) 4876-8259
E-mail: lnm@springer.com